Applications of
Ambient Energy in Buildings

Applications of Ambient Energy in Buildings

edited by

A. F. C. Sherratt

BSc PhD CEng FIMechE FCIBS FInstR MASHRAE
Assistant Director, Thames Polytechnic, London, UK
Editor International Journal of Ambient Energy

London   New York
E. & F. N. Spon

First published 1983 by
E. & F. N. Spon Ltd
11 New Fetter Lane, London EC4P 4EE
Published in the USA by
E. & F. N. Spon
733 Third Avenue, New York NY10017

© 1983 Construction Industry Conference Centre Limited,
PO Box 31, Welwyn AL6 0XA, UK
Chapters 2 and 5 © Crown Copyright
Chapter 9 © P. O'Sullivan and P. J. Jonas

Printed in Great Britain by
J. W. Arrowsmith Ltd., Bristol

ISBN 0 419 12790 9

British Library Cataloguing in Publication Data

Applications of ambient energy in buildings.
  1. Building   2. Architecture
  3. Renewable energy sources
  I. Sherratt, A.F.C.
  721   TH153
  ISBN 0-419-12790-9

Library of Congress Cataloging in Publication Data

Applications of ambient energy in buildings.

Based on the second 'Ambient Energy in Buildings'
conference organized in March 1981 by the Construction
Industry Conference Centre Ltd and others.
Bibliography: p.
Includes index.
  1. Buildings—Environmental engineering—Addresses,
essays, lectures. 2. Renewable energy sources—Addresses,
essays, lectures. 3. Buildings—Power supply—Addresses,
essays, lectures. I. Sherratt, A. F. C. II. Construction
Industry Conference Centre.
  TH6025.A66   1983   690   83-14766
  ISBN 0-419-12790-9

Contents

Preface

Ambient energy might be defined literally as 'the energy that surrounds us'. Solar energy and wind energy are the forms of ambient energy most likely to be related directly to buildings. Other forms – tidal, wind, geothermal and biomass – are likely to have application indirectly.

Few people will need to be reminded that resources of fossil fuel are finite and in the most convenient forms – oil and natural gas – will last for a relatively short time compared to the length of human lives. Our standard of living is energy based and if present day quality of life is to be maintained in the developed world and, perhaps more important, the peoples of the poorer countries of the third world brought towards that standard, ways to eke out fossil fuel reserves are needed. Alternatives must be found to save fossil fuel for the more vital tasks.

What is the right alternative is perhaps open to debate. There is little doubt however that to minimize energy requirements and maximize the use of ambient energy cannot be bad provided it can be shown that it is or would be cost-effective.

This book, which is based on the proceedings of the second 'Ambient Energy in Buildings' conference, examines the technology and economics of the application of ambient energy – predominantly solar energy – in a variety of buildings. It assesses the current stage of developments, documents examples – both successes and failures – and documents much ambient energy experience relevant to the UK, Europe and other parts of the world with similar climates.

In the assessment of ambient energy systems the ideas and methods of conventional economics are used. At the end of the book John Twidell in a hard-hitting, thought provoking, well-reasoned chapter, develops a quite different way of looking at ambient energy projects introducing a new concept of 'resource ethics'. I suspect that, read carefully, this chapter will convince most people that there are considerable flaws in our current perspective on ambient energy and that Dr Twidell's suggestions warrant deep consideration.

The opinions expressed in this book are the authors' and not necessarily those of the editor, publisher or the Construction Industry Conference Centre Limited.

A. F. C. SHERRATT

It is with sincere regret that I record the death of Peter Jonas in September 1982 after a long illness: A tragic loss to his many friends and colleagues and to the great debate on energy.

A.F.C.S.

Acknowledgements

The editor and organizing committee wish to thank all the people who have participated in the production of this book and in the arrangement and operation of the conference on which it has been based.

Special thanks are given to Jean Stephens, Diana Bell and their colleagues at the Construction Industry Conference Centre for their courtesy and efficiency in the organization and administration.

Organizing Committee and Contributors

This book is based on the second 'Ambient Energy in Buildings' conference organized in March 1981 by the Construction Industry Conference Centre Ltd in conjunction with the Chartered Institution of Building Services, the Royal Institution of Chartered Surveyors, the Institute of Energy, the Department of the Environment and the Department of Energy.

ORGANIZING COMMITTEE

Chairman: A. F. C. Sherratt, BSc, PhD, CEng, FIMechE, FCIBS, FInstR, MASHRAE

D. Fisk, MA, PhD
(representing the Department of the Environment and the Chartered Institution of Building Services)

R. Jackson, MSc, FInstP, FIEE, FIM, AIMC, SFInstE
(representing the Institute of Energy)

P. J. Jonas, BSc(Tech), BSc(Econ), MIEE, MIMechE
(representing the Department of Energy)

J. C. McVeigh, MA, MSc, PhD, CEng, FIMechE, MIProdE, MCIBS
(representing the Chartered Institution of Building Services)

A. E. Turner, FRICS
(representing the Royal Institution of Chartered Surveyors)

CONTRIBUTORS

D. Allen FRICS
Partner (Quantity Surveying), Building Design Partnership, Manchester, UK

Andrew Burke MA DipArch RIBA
South London Consortium, London, UK

J. Campbell MCIBS
Technical Director, Ove Arup Partnership, London, UK

M. Corcoran MSc MCIBS
Associate, Building Design Partnership, Manchester, UK

P. D. Dunn BSc PhD FCIBS HonFRIBA MIOA
Professor of Engineering Science, University of Reading, UK

Ivan Dunstan BSc PhD CChem FRSC
Director, Building Research Establishment, Garston, UK

John R. J. Ellis CEng MIMechE MCIBS
Partner *(Engineering Services), Building Design Partnership, Manchester, UK*

H. Hörster Dr rer nat
*Deputy Director, Philips GmbH Forschungslaboratorium, Aachen, West Germany
Chairman, German Section, International Solar Energy Society*

P. J. Jonas BSc(Tech) BSc(Econ) CEng MIEE MIMechE
Head, Energy Conservation Technology Branch, Department of Energy, UK

Julian Keable FRIBA AADipl
Senior Partner, HELIX Multi Professional Services, Reading, UK

Anthony Kirk DipArch RIBA
Consultant, 1 Grove Close, Hayes, Kent

P. O'Sullivan BSc PhD FCIBS HonFRIBA MIOA
*Professor of Architectural Science, The Welsh School of Architecture, UWIST,
Cardiff, UK*

Robert W. Todd BSc(Eng) PhD
Technical Director, National Centre for Alternative Energy, Machynlleth, UK

John Twidell MA DPhil
*Lecturer, Department of Applied Physics, and Energy Studies Unit, University of
Strathclyde, UK*

John Willoughby BSc MPhil MCIBS
*Energy Consultant and Associate Senior Lecturer, Gloucestershire College of Arts
and Technology, Gloucester, UK*

S. J. Wozniak BA PhD
*Head of Solar Technologies Section, Building Research Establishment, Garston,
UK*

1

Ambient Energy and the Environment in Buildings

Ivan Dunstan

INTRODUCTION

A good deal has happened in the UK since the ambient energy and building design conference held in 1977 [1]. Research interest has grown, the volume of publication has increased significantly, development and design studies have been undertaken, and a number of low-energy buildings have appeared incorporating ambient-energy features.

Some important conferences and symposia have taken place nationally and internationally, and 1979 was a year of some note with CIB's second international symposium, 'Energy conservation in the built environment', in Copenhagen [2], when there was a valuable exchange of views on the cost effectiveness of energy conservation measures; and the major RIBA conference in London in September, 'Buildings – the key to energy conservation', [3] which drew together many interrelated developments, and led to the publication of a series of fifty case studies of low-energy buildings erected or under erection in the UK. That review covered a wide range of building types – houses, offices, schools, covered swimming pools, factories, hospitals, a warehouse, and a church. Later in this chapter I shall mention one of these case studies – a low-energy office which is currently being commissioned on the BRE site at Garston.

This book provides a splendid opportunity to review progress in a specific, but very important, aspect of energy in relation to buildings – namely the impact of research and development on the application of ambient energy to the building itself.

Over the last three years there have been some changes of emphasis. Although

Author's note: The views expressed in this chapter are those of the author and not necessarily those of the Building Research Establishment.

the autonomous or zero energy building is an attractive and challenging design idea, the increasing need to evaluate options that cover large parts of the building stock has concentrated more attention on less ambitious design objectives. The autonomous house may well have a role in areas remote from conventional energy supplies but it seems likely that for the majority of the UK building stock, some of the most promising ambient-energy sources would meet its needs not at site but through the traditional energy-supply grids. What is certainly true, moreover, is that a greater awareness has developed of the possibilities for meeting the basic energy requirement of buildings by direct use of ambient energy in the form of light, heat, and wind. It is an approach which complements the high-technology solutions offered by aerogenerators, photovoltaic cells and the use of wave and tidal power.

In this chapter I shall not be concerned with the use of ambient energy as a source of supply to the energy grid system in the UK, though clearly a great deal of research is aimed towards that objective. There will be no mention therefore of multiple arrays of solar panels or photovoltaic cells; nor will there be any reference to tidal power, geothermal energy, or windmills as high as Big Ben.

The chapter *is* concerned with developments of a kind which are closely associated with the design, the fabric and the services of the *building* rather than with the supply grid. The scope is still wide; it embraces active and passive solar heating, daylighting, natural ventilation, and, of course, heat pumps.

The main aim of the chapter and its common theme is to explore just how far and in what manner the research and development is working its way through into practice, because it is this which in the end will determine its ultimate acceptance and success. This, of course, is the classic route through research and development and into design, specification and production which is well defined and understood in manufacturing industry. It is tempting to draw a parallel between these familiar stages characteristic of manufactured products, and the design and construction process applied to building. There are similarities, but closer examination reveals very many differences due to numerous factors, not least the nature of the building and construction industry with its many sectoral interests, its multitude of small firms, and its perception of research; then too there is the building itself which cannot be treated as a collection of unrelated elements – it is a system with interdependent sub-systems, and it has to be treated as such.

Thus, the usual criteria which apply to research and innovation – Is it desirable? Is it feasible? Is it acceptable? – have more than their usual significance when applied to building and construction. The way has to be prepared very carefully, the course charted, to ensure that the new and unfamiliar is properly tried and tested, that its introduction will not prejudice the performance, durability, or cost of the building in some unexpected way, and that those responsible for the process of building understand the critical features of the innovation – its robustness during building or in use – so that error is minimized and expectations are realized.

Mistakes can be expensive; for example, over the years, the housing stock in the UK – amounting to about 20 million dwellings – is replaced at the rate of about 1% per annum, some 200 000 dwellings each year. The scope for successful innovation in energy conservation is immense, and there are commensurate benefits; however, by the same token, ill-advised innovation on any scale could have disastrous consequences – and adverse effects do not necessarily appear in the short term. It is for this reason that the chapter emphasizes the importance of validating research and development by demonstration, the importance of supporting design with acceptable standards, and the importance of encouraging good building practice by attending to what actually happens on the building site.

I mentioned that the scope for successful innovation in energy conservation in buildings is immense. May I illustrate that.

Figure 1.1, taken from Department of Energy paper No. 39 (*Energy Technologies for the United Kingdom*) [4] shows the world primary energy consumption in millions of tons of coal equivalent so far this century, in terms of the overall supply, and the individual fuels.

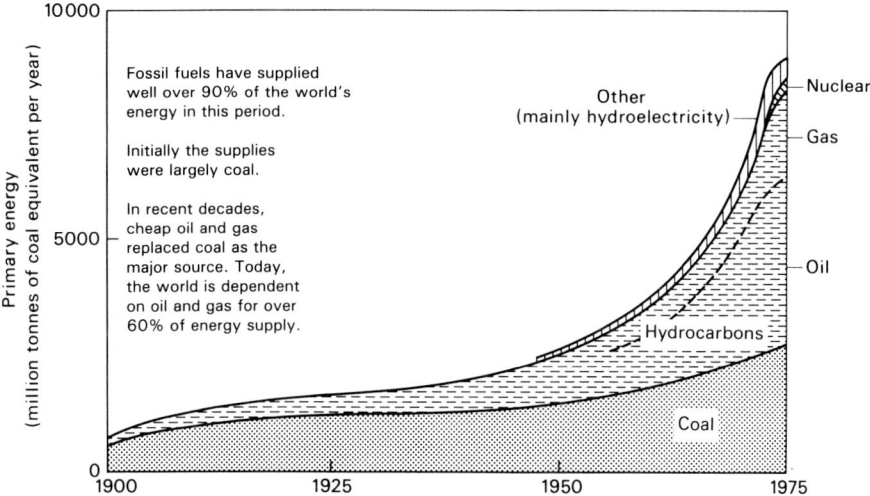

Figure 1.1. World energy supplies – the 75 years 1900–1975

Looking ahead to the middle of the next century, Fig. 1.2 shows a projected world energy demand for the next 75 years – consistent with the low energy growth case (barely doubling between 2000 and 2050) put to the 1977 World Energy Conference. The demand for energy is likely to be driven both by the growth in the economies of the industrialized world, and increasingly by the growing energy needs of developing countries.

There is an important consequence. The world population is also expected to grow over the same period and the current distribution of energy usage – with the

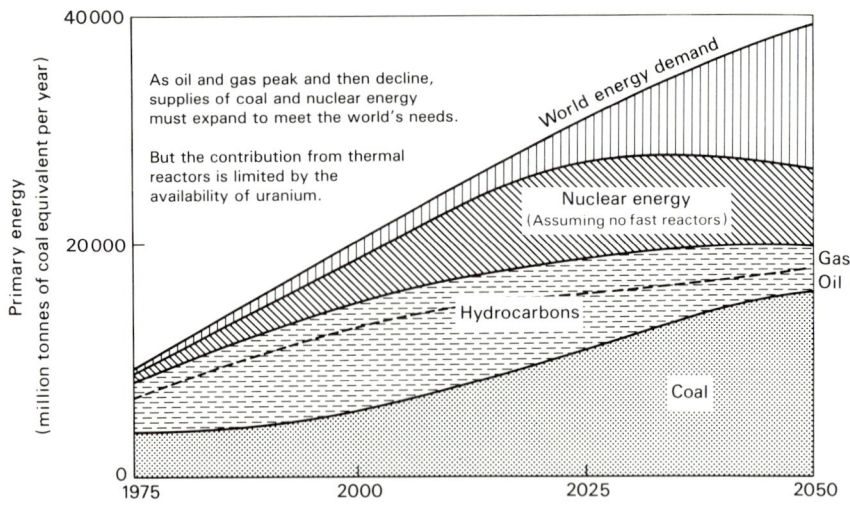

Figure 1.2. World energy requirements – the 75 years 1975–2050

Western industrialized world representing less than 15% of the world's population consuming more than half of the total energy demand – shifting towards developing countries. Thus the per capita consumption of today's major consumers must fall, if more developing countries become industrialized and energy supply follows this projection.

Turning now to the position in the UK, Fig. 1.3 shows the energy balance for 1979 comparing, on a primary energy basis, the fuel consumed (9.3×10^9 GJ) with the energy consumed. The primary energy consumed is always greater than the consumption by the user, the difference arising from the energy expended in processing and in distribution before the fuels reach the building as delivered energy.

The proportion of energy consumed annually by building services is not known accurately but it is generally accepted that it is about a half of all primary energy. Figure 1.4 shows that the domestic sector is the largest single sector accounting for 29% of the primary energy consumption. A further 20% or so is consumed in building services in industry – for heating and lighting factories – and by the other users sector which includes schools, hospitals, and public buildings.

It is this massive consumption of energy in buildings – estimated to cost the nation about £10 000 million per annum – which is the prime target for energy conservation measures, and for the contribution ambient energy can make to those measures. In the long run it has been suggested that the primary energy consumption of buildings could be decreased by 25–30% of its present value with no loss in environmental quality or comfort.

Against this background I would like to comment on each of the areas where

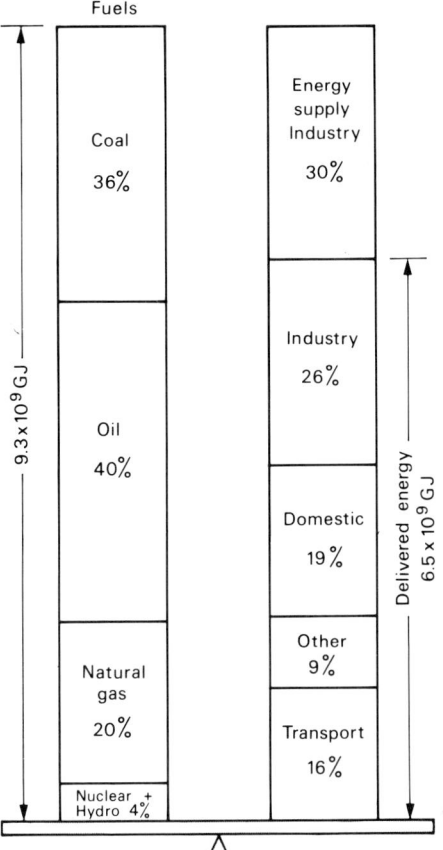

Figure 1.3. Energy balance for UK (1979) including consumption by energy supply industry

ambient energy can make its contribution in the particular context of the environment in buildings.

SOLAR HOT-WATER HEATING

The use of a solar collector for heating domestic hot water has been a prominent feature in many designs seeking to exploit ambient energy. They can, however, be added to existing buildings, and in many exercises solar collectors have been used to supply part of the water-heating load, with varying degrees of success. The option is not cheap; recent costs to BRE indicate that a well-installed 4.5 m² solar collector system will cost a householder in the region of £1000. It is interesting to note that the greater part of that cost is in fact for materials and skills which can

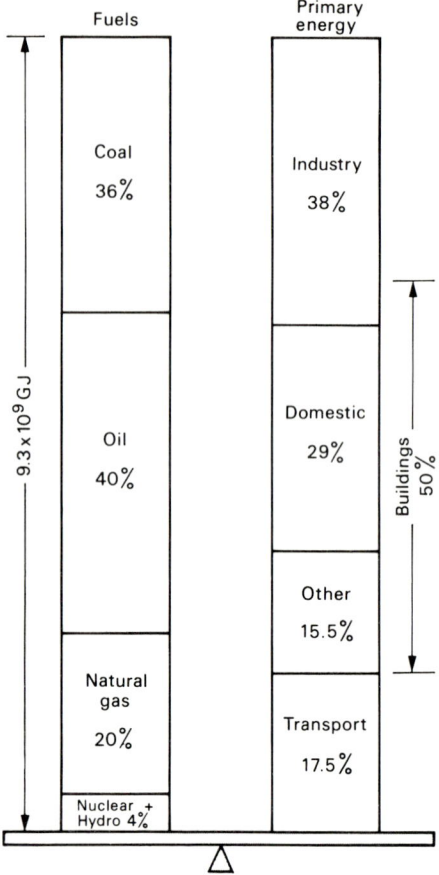

Figure 1.4. Primary energy balance for UK (1979) showing consumption by buildings

be considered as traditional building and building services work. Although it is always difficult to judge the relationship between cost and rate of return that defines the market share of an energy conservation product, at this level of investment it seems likely that many purchasers will desire a predictable level of performance. At this point in the development of the UK solar industry adherence to standards and other recognized technical guidance may be expected to play an important part in gaining widespread customer confidence. For this reason a great deal of research effort has been undertaken at BRE and elsewhere on the development of the new British Standard for domestic solar water heating [5]. Publication of that Standard must count as one of the major milestones in the development of ambient-energy sources in recent years. The designer/installer and potential owner also now have access to far more detailed technical guidance. For example many of our own experiences, some based on our laboratory and field trials, are presented in a recent HMSO book on the subject [6].

SPACE-HEATING SOLAR COLLECTORS

Although a number of design studies have been conducted in this area few projects have been taken through to the construction stage. Two of the solar-heated houses at BRE, both supported by heat-pump technology, were described at the last Ambient Energy and Building Design Conference. One of those uses an air collector, and the other uses a water collector and several water thermal stores. The work undertaken by John Laing R&D and the Calor Group has resulted in the marketing of a number of sophisticated, active solar-heated houses. A feature of some of these dwellings is the use of a new form of thermal storage which may assist more widespread application of active solar space heating in the UK climate. However, successful application of new types of low-temperature storage even at reasonable cost, does not by itself guarantee a sizeable UK market for solar space heating. Many other ambient energy technologies also perform better at large load factors. It therefore remains to be seen, as further experience is gained of the realized cost in construction and long-term performance, how competitive this type of space heating will be.

PASSIVE SOLAR HEATING

I would like to turn now to passive solar heating – usually taken to mean the heating which arises from radiation falling on roofs or walls, or transmitted through windows of buildings. Much depends of course on the availability of that radiation and it is not surprising to find that the benefits of passive solar gain have been explored most enthusiastically in regions of the world where the incidence of sunshine can be guaranteed with a fair degree of certainty.

The sun-lit scene in Fig. 1.5 must be California, and it shows a water wall – a wall composed literally of drums of water which absorb heat during the day and release it to the building during the night. The containers, which are on the south-facing wall, are painted black, and are covered with insulating shutters at night.

This may not be an optimum design solution for the UK in its simplest form because the net heat balance of the system may be unfavourable during most of the heating season. However, some of the basic principles of designing for passive solar energy gains are being incorporated into UK buildings, and we have one example close to home in the new low-energy office block which has just been built at Garston [7]. The building which is shown in Fig. 1.6, was designed by the Property Services Agency in conjunction with BRE staff and it will be used chiefly to run the BRE research programmes on the use of microprocessors for the control of building services.

Ambient energy is used in two respects: there are solar panels to provide hot water for washbasins; and the fenestration has been optimized on the north- and south-facing facades of the building. Although the north and south facades look superficially identical there is, in fact, a considerable difference in window area.

Figure 1.5. A water wall in California

Figure 1.6. BRE low-energy office block

Those on the north are 30% of the total facade, whereas those on the south are 50%. This relatively large window area on the south-facing facade may, of course, result in overheating during summer months, and the experimental design therefore incorporates external sun-blinds operated by a central solar sensor.

An important feature of this low-energy office block is that it is adjacent to a conventional office block of similar dimensions. This will enable us to compare the performance of the two buildings in a more or less identical climatic environment.

The design target for energy usage in the low-energy office block is 50% of the energy used in a similar, conventional office, and 35% of that used in a corresponding air-conditioned building.

DAYLIGHTING AND NATURAL VENTILATION

There are two well-established features of building design which merit a new look in the context of energy conservation and the use of ambient energy. These two features are daylighting and natural ventilation.

It is self-evident that daylight is 'free' energy whereas artificial lighting adds to the energy bill – and the cost can be high in, for example, large office blocks. Savings can undoubtedly be achieved both by optimizing the size of windows and by better control of artificial lighting.

I have taken an example from BRE's recent research in a school building near

Figure 1.7. School lighting experiment

Reading. Figure 1.7 shows a deep-plan room in the school with three banks of lights. The outer bank, those lights nearest to the windows, are controlled on an on/off basis photoelectrically; the centre bank is also controlled photoelectrically and provides continuous dimming or top-up; and the inner bank is on permanently when the room is occupied. This arrangement saved 30% of the energy originally used for lighting the room.

Following a similar argument, natural ventilation can be regarded as a source of 'free' energy to off-set energy demands for the provision of fresh air or cooling by other means [8].

Both aspects – daylighting and natural ventilation – are brought together in a series of environmental design aids prepared by BRE. These can be used by the architect at an early stage of the design process to determine whether a building can be daylit and naturally ventilated – with consequent energy-saving – or whether it requires daytime artificial lighting, and mechanical ventilation, possibly with cooling.

Figure 1.8 shows a combined thermal and wind ventilation design aid to examine the environmental conditions in a room with windows protected by an external Venetian blind. The horizontal scale shows window size, and the vertical scale shows daytime ventilation rates. Relevant parameters are the mean indoor temperature (22.5°C, 23°C, 25°C), the indoor temperature *range* (5°C, 4.5°C, 4°C, 3°C) and the percentage of glazing which can be opened (70%, 50%, 30%). The unshaded area represents an acceptable degree of thermal comfort, and this provides the design parameters. With a less effective form of sun control it is likely that the required ventilation rate could not be obtained by natural means.

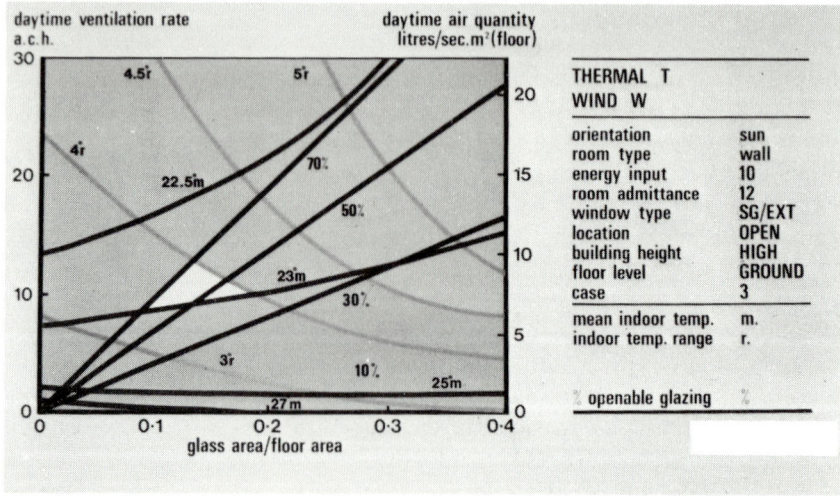

Figure 1.8. BRE thermal and wind ventilation design aid

HEAT PUMPS

The technology of using heat pumps is already widely recognized within the commercial building sector, as one of the options aimed towards building heat recovery. However, it is also an important option in taking heat from the ambient, either the air or another convenient source such as the ground. A further option is to use the heat pump in conjunction with another low-temperature heat source, such as a solar collector, or a thermal store, to provide heat at a required temperature. Heat pumps figured in the previous ambient energy conference, and since that time there has been an opportunity to review their contribution to low-energy building technology. Exercises have now been undertaken ranging from the small, single-room heat pump through small-scale domestic heating applications up to moderate-scale group heating schemes. The heat pump market has continued to develop in continental Europe, and a number of UK manufacturers have now begun to offer products in the market place. Although there are great promises for cost reductions in this technology there is a strong need for reliable performance standards, as indicated by the wide variability in performance obtained, often at variance to that indicated by a simple interpretation of manufacturers literature. In this respect the United States has taken an important lead and the work at the National Bureau of Standards is well towards defining the type of performance standard required of their heat pumps. One of the heat-pump trials in the BRE low-energy houses has certainly shown that heat pumps are able to give a running cost performance close to that of a well-designed, natural-gas, central-heating system [9]. Work is still required to establish the likely future cost reductions, both in production and in installation, in new and existing buildings.

THERMAL INSULATION

Thermal insulation of buildings continues to be an essential element of energy-conservation measures. Changes have taken place in the UK Building Regulations which, together with a reduction of wintertime heat losses due to conduction through the fabric, will increase the utilization of solar energy. On 1 June 1979 new building regulations for non-domestic buildings came into force which require an average U-value of 0.6–0.7 W/m^2 °C for the opaque parts of walls and roofs, together with limits on the area of glazing. High levels of insulation can reduce the length of the heating season since solar energy and miscellaneous heat gains provide the necessary heat towards each end of the season [10]. This trend towards higher levels of insulation is likely to continue as proposals for applying similar levels of insulation to houses are now being considered.

BUILDABILITY

Attention thus far has been directed towards the options which at present appear to be developing as the strong contenders for ambient energy use in building design, when applied across the entire building stock. The discussion has also served to indicate the many stages beyond the drawing board that are required to be established before a particular option can secure itself as common-day practice. Standards have been seen to play an important role, but equally is a realization of what can practically be achieved in the construction process.

For several years now a small group of researchers at BRE has been examining the incidence of faults in building, and I would like to illustrate some of their findings, using a series of pie diagrams. The data are the result of very many observations at fifteen different house-building sites.

First, two diagrams showing the origin of faults: Fig. 1.9 based on faults by number of kinds, and Fig. 1.10 based on their estimated frequency. In each case a substantial proportion of the faults could be related to what happened on site – generally because of a lack of care rather than lack of knowledge. But many faults are attributable to the design stage of the process.

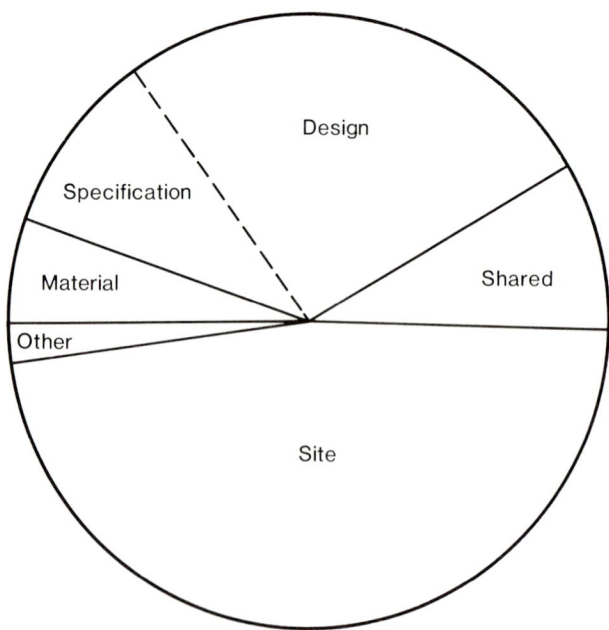

Shared = Design, but with some site responsibility.
Faults of this kind included in design and
specification in Figure 1.10

Figure 1.9. Origin of kinds of faults

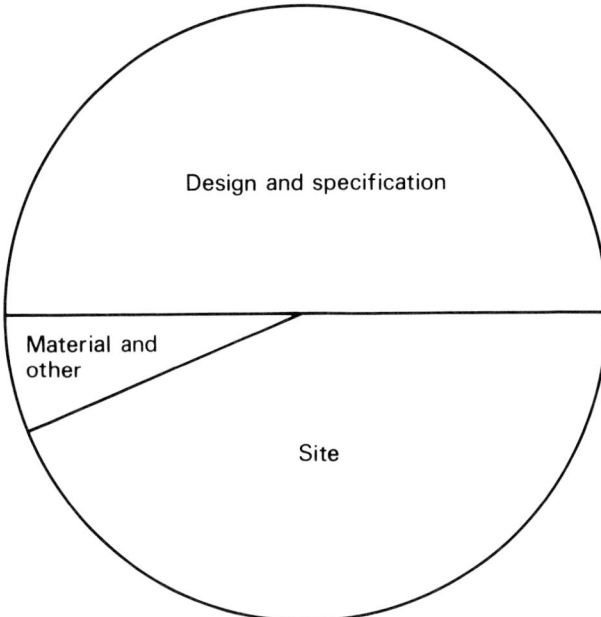

Figure 1.10. Origin of faults taking frequency of repetition into account

In Fig. 1.11 the faults are analysed in relation to building performance. Sectors of interest are those labelled thermal insulation (4.7%) and ventilation/heating/condensation (7.6%). The last pie diagram (Fig. 1.12) summarizes results for defects arising from innovation. Once again workmanship and design predominate as causal factors; but it is interesting that materials can also contribute to the problem in a more significant way when the building involves some innovative feature.

This work is published in full elsewhere; my purpose in quoting some of the results here is to emphasize the importance of ensuring that buildability is taken into account when changes are introduced.

CONCLUSION

In many respects progress in research and development on ambient energy and building design is following a well-established pattern – basic studies leading to innovation in equipment and design; innovative measures being underwritten through standards, design aids, and codes of practice; innovation being introduced into practice; and lessons learned being fed back to research and design.

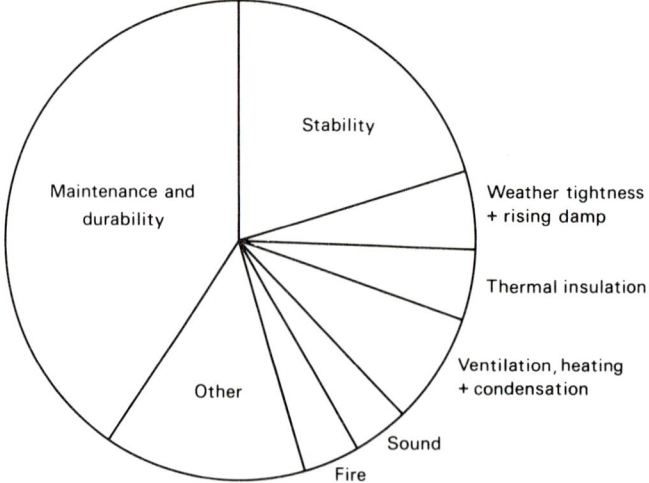

Figure 1.11. Attribution of all faults by performance

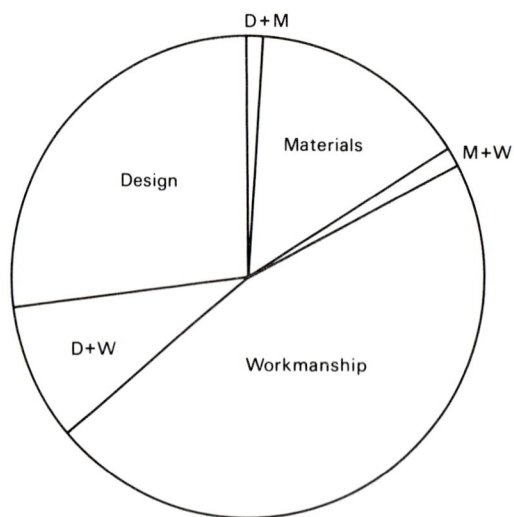

Figure 1.12. Origin of faults associated with innovation

Progression has been steady rather than dramatic, évolutionary rather than revolutionary, but significant in terms of basic understanding of principles and appreciation of practicalities. It is a pattern of progress which fits well with the overall aim of achieving reliable, cost-effective solutions to the problem of energy supply and utilization in buildings.

ACKNOWLEDGEMENT

It is a pleasure to acknowledge my indebtedness to colleagues at BRE, particularly Dr D. J. Fisk, H. W. Harrison, and Dr S. J. Wozniak, for their advice and assistance in the preparation of this paper.

REFERENCES

1. Randell, J. E. (ed.) (1978) *Ambient Energy and Building Design*, Construction Press, London. (Based on conference papers of the first CICC Ambient Energy and Building Design Conference, April 1977.)
2. CIB (1979) *Second International CIB Symposium on Energy Conservation in the Built Environment*, Danish Building Research Institute, Copenhagen, Denmark.
3. Kasabov, G. (1979) *Buildings – the Key to Energy Conservation*, RIBA Energy Group, Royal Institute of British Architects, London, p. 96.
4. Bonshor, R. B. and Harrison, H. W. (1982) *Quality in Traditional Housing, Vol. 1 An investigation into faults and their avoidance*, HMSO, London.
5. British Standards Institution (1980) *Code of Practice for Solar Water Heating Systems for Domestic Hot Water*, BS 5918, BSI, London.
6. Wozniak, S. J. (1979) *Solar Heating Systems for the UK*, HMSO, London.
7. Uglow, C. E. and Petherbridge, P. (1981) The BRS low energy office building – aspects of a design study paying attention to the passive use of solar energy. *Building Services and Environmental Engineer*, **6**(3), 6–9.
8. Warren, R. P. and Webb, B. C. (1980) Ventilation measurements in housing. In *Proceedings of the CIBS Symposium 'Natural Ventilation by Design'*, Garston, December 1980, Chartered Institute of Building Services, Balham.
9. Mountford, D. and Freund, P. (1982) The field performance of a well-insulated house heated by an air–water heat pump. *Int. J. Ambient Energy*, **3**(1), 9–18.
10. Siviour, J. B. (1976) Designs for low energy houses. In *Proceedings of the 1976 International Symposium of the CIB*, British Building Research Establishment, Garston, UK.

DISCUSSION OF CHAPTER 1

Dr J. C. McVeigh (Brighton Polytechnic). Dr Dunstan stated that the demand for energy will increase in the industrialized countries as their economies grow. This is the 'conventional wisdom' approach often quoted by the Department of Energy in energy forecasts. More recent studies show that this conventional view based on figures of the 1960s and early 1970s is no longer necessarily valid. There is ample recent evidence (some twenty authoritative studies) to suggest that economic growth is perfectly possible without growth in energy consumption. In fact, some of the buildings described in the chapters of this book illustrate admirably that possibilities for energy conservation exist which have little or nothing to do with growth in the economy. The economy could be growing, with industry doing more work and at the same time reducing the overall use of energy. Any suggestion that a growth in the economy means a growth in the energy use must be treated with considerable caution. Statistics for the OECD countries show that the decoupling of economic growth from energy is already occurring.

Dr S. J. Wozniak (Building Research Establishment). The diagrams used by Dr Dunstan were taken from Energy Paper 39.* It is recognized that the future of energy supply is uncertain and this is why research will continue over a wide spectrum of technologies. The prospects for each technology will be continually reviewed to help ensure that priority is given to developing and proving cost-effective systems. This should help facilitate application of an appropriate mix of technologies on a wide scale.

Industrialization of an under-developed country would inevitably result in an increase in energy consumption within that country. However, I believe it is widely accepted now that there is considerable scope for reducing energy consumption in the presently developed countries without much if any reduction in material living standards.

There will no doubt continue to be considerable debate as to how best to invest in energy conservation. From a national point of view reduction in the rate of consumption of fossil-fuel reserves is an important factor, whereas some individuals may attach greater importance to use of solar energy *per se.* Cost-effectiveness is likely to remain a good indicator of total resource usage.

*Energy Paper 39 (1979) Energy Technologies for the United Kingdom: An Appraisal for R, D and D Planning. Department of Energy, HMSO, London.

2

Assessment and Use of Solar Collector Systems in the UK

S. J. Wozniak

APPLICATION OF SOLAR HEATING SYSTEMS: A BRIEF HISTORY

Solar radiation data collected by UK meteorological stations has been available for decades. The energy incident annually upon 1 m² of horizontal surface in the UK at present levels of air pollution is about 900 kW h and can exceed 1000 kW h on a suitably inclined, south-facing surface in favoured areas. These figures set obvious maxima to the likely output of systems designed principally to capture solar radiation.

The pioneering work of Heywood showed that although collection efficiencies for solar water heating of greater than 60% could be attained on good days using conventional flat-plate collectors, an overall annual efficiency of 35–40% was unlikely to be exceeded in the UK [1, 2]. In 1975 BRE published an appraisal of solar water heating which drew on both UK and overseas experience. It was concluded that whilst conventional solar collector systems had been shown to achieve 45–55% efficiency under South African conditions a likely annual efficiency for these systems in the UK was 35% [3]. Similar figures were published by the then newly formed UK branch of the International Solar Energy Society.

During the five years following 1973 application of solar energy systems in the UK was primarily for domestic hot-water preheating, and collector areas ranged from less than 2 m² to (occasionally) over 5 m² per house. Some of the systems were poorly engineered and were sold using dubious sales techniques and literature. These aspects of the solar industry received publicity as a result of prosecutions brought under Trading Standards legislation. Typical of the claims made over this period were that more energy could be delivered from systems than was incident upon their collectors and that the systems would be long-lasting and maintenance free. Parallel developments occurred in other European countries and highlighted the need for recognized test procedures, performance data and guides to good technical practice to assist a healthy development of the solar industry.

Despite a poor solar climate and (until recently) relatively low fuel prices popular interest in solar technology in the UK has been considerable. The relatively slow growth of the UK industry compared with that in some overseas countries, particularly the USA, has had the useful result that design errors in some early solar water-heating systems have not been repeated in more than a few thousand dwellings and this has served to limit adverse publicity.

Following the events of 1973 a Department of the Environment programme to study solar collector systems was set up at BRE in 1974. A short time later a larger Department of Energy programme based mainly on extramural contract work with industry, universities and consultants was formed following the ETSU report on the potential for solar energy in the UK [4]. The object is to assess the likely size and timing of the contribution of solar energy to the UK energy supply, which at the present is still the subject of some uncertainty.

Cooperation between government establishments, universities, private research organizations and trade associations resulted in publication of two consensus guides to good practice that give numerical design and engineering detail for small solar systems. The HVCA guide [5] published in 1979 was followed early in 1980 by the BSI code of practice [6] and both have become accepted as useful working documents. The HVCA guide is intended primarily for contractors and their customers and sets out in simple terms not only the basic engineering do's and don'ts but also the cost effectiveness of solar systems. By contrast the BSI code gives the framework of performance prediction methods that are likely to become universally accepted in the UK but contains less plumbing detail and no study of cost effectiveness.

Early BRE work [7] resulted in publication of computer predictions for solar water-heating systems having a range of characteristics. The results of a more comprehensive study are given in the British Standard. Although the subject of some contention whilst in preparation, inclusion of these results in the BSI code can now be regarded as a milestone in the application of solar heating in the UK. Briefly, it is shown that the performance of domestic systems is likely to be determined largely by the collector characteristics, the collector area, and the average daily hot-water demand. The importance of secondary factors such as pipework insulation is also quantitatively estimated. It is shown that the orientation and slope of the collector may have less effect on the annual thermal performance than might be supposed. With some reservations these results can be used to predict the output from large systems where these are upscaled versions of those studied. Where the system characteristics are significantly different, however, neither the predicted outputs nor the BS collector classification system may be used. For the types of collector sold in the UK, many of which are thought to lie in class IV or V, the solar energy supplied annually when used in a typical domestic hot water system will lie in the range 1 ± 0.3 GJ/m^2. A

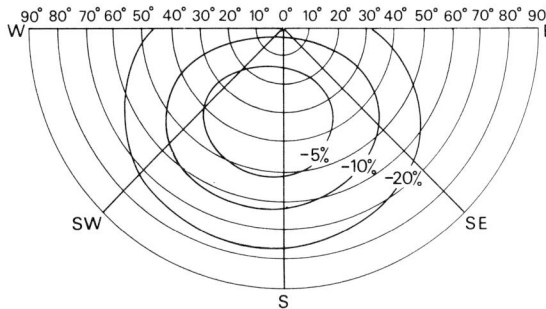

Figure 2.1. Variation of the solar energy supplied by the reference system with a class IV collector, with orientation and tilt of the collector (reproduced from BS 5918 by permission of the British Standards Institution)

class IV collector (as defined in BS 5918[6]) could be either double glazed with a neutral black absorber or single glazed with a good selective surface; collectors of lower performance are grouped in class V. Figure 2.1 shows the predicted variation of the solar energy supplied for a given type and size of system with orientation and tilt of the collector. It can be seen that for tilt angles between 10 and 45° and orientations between SE and SW there is less than 10% reduction in system output compared with the optimum case. This is an encouraging result since it implies that many UK houses may have roof slopes suitable for solar collector systems. Less encouraging but in conflict with neither the experimental work of Sharpley [8] nor more recent BRE work are the system outputs.

One deficiency of both the HVCA and BSI documents is that they fail to give detailed technical guidance in some of the areas where it is required, and the BSI code in particular has been criticized in this respect. However, both are consensus documents prepared against a background of considerable discussion in the industry and inclusion of additional detail on contentious topics would have further delayed their publication.

In 1979 BRE published a book dealing principally with the engineering of small solar systems [9]. Other topics such as swimming-pool and space-heating systems are also covered. Building-oriented problems such as fixing solar collectors to roofs that are glossed over in other texts are discussed in detail, with some of the ideas being taken from supervision of field trials in which it was necessary to design completely new and, it is to be hoped, satisfactory mounting systems.

Seven different types of primary circuits representing in essence the current practice at that time are described and their advantages and disadvantages are outlined. Types 1 and 2 (thermosiphon systems) are still not widely used in the UK although there is now increased interest in their adoption. All the other system types (except perhaps type 4) are still being used with only minor variations.

One of the most noticeable if not most significant technical developments in domestic solar system design in the last year or so has been the increased use of multi-function electronic control systems linked not only to solar circuits but also to flue dampers, room and water thermostats, motorized valves and boiler controls. The complexity of some of these systems is such that reliable assessment of their impact on total fuel consumption would have to be made on an individual house basis.

There is room for development both in matching control-system design to real requirements and in heating controls that provide only for a predetermined 'background' temperature and which need intervention by the user every time a higher temperature is desired. One of the first uses of such a control system is being investigated as part of a BRE experiment using a heat pump and solar collectors to supply twelve houses with space and water heating [10].

The advantages of this approach for water heating are threefold and have already been discussed in detail [11]: solar systems will work more efficiently because of the greater volume throughput, heat losses will be lower (assuming no increase in cylinder size) and there may be a reduction in the overall heat requirement. This latter advantage is speculative but given that some uses of hot water may be for a given time (rinsing of hands) not unreasonable. Hidden within these ideas is the tacit supposition that existing standards of domestic hot-water supply may originate more in the historical use of oversized heating equipment than in a rational appraisal of likely demand in an age of increasing energy prices. In addition it has to be assumed that energy prices will increase to a level at which they become a significant burden for both households and institutions, thus helping to ensure co-operation in reducing consumption.

Looking to the future there is another area that will need discussion. The performance of flat-plate collector systems that supply both space and water heating in well-insulated dwellings in the UK can be markedly influenced by the summertime consumption of hot water. In simple terms, the space-heating season may be short and restricted largely to those months in which the incidence of solar radiation is lowest. The total annual space-heating requirement may be some 30 GJ of which (it will be supposed) a solar system might supply 10 GJ. For this to occur during the months October–March without long-term storage would require about 30 m^2 of collector operating at 30% efficiency. If the conservationist future is seen as one in which consumption generally is restricted to necessity rather than to extravagance it is not unreasonable to assume a four-person household water usage of about 150 l per day. If this figure were to be doubled then making the assumption that the collectors could supply all the additional energy requirement over four months and half of it over another two months the solar contribution increases by some 4.5 GJ per year.

If the heating season were to be drastically shortened [12] perhaps to the period November–February the space-heating contribution falls to less than 5 GJ and the significance of the high summertime use of hot water becomes clearer. The essential point for discussion is whether present patterns of hot water consumption are likely to change in a future in which greater use of solar systems is encouraged by high energy prices or restricted fuel availability. Reduced consumption has already been postulated as one mechanism that may limit the usefulness of small domestic systems [13].

THERMAL PERFORMANCE EVALUATION OF COLLECTORS

Assessment of the thermal characteristics of collectors has been carried out on a small scale for many years in countries having favourable climates. Often, system tests rather than pure collector tests have been used. One of the consequences of the recent upsurge of interest in solar heating was the publication in the USA of two test methods for rating collector performance [14, 15]. Both these methods require extended periods of constant high-intensity sunshine during which the collector may be brought to a steady state. Each such state yields one point on the 'collector characteristic' or, more colloquially, the 'NBS curve'. In reality since the overall heat loss coefficient of most collectors is dependent on temperature a more accurate representation is possible using a family of such curves, each

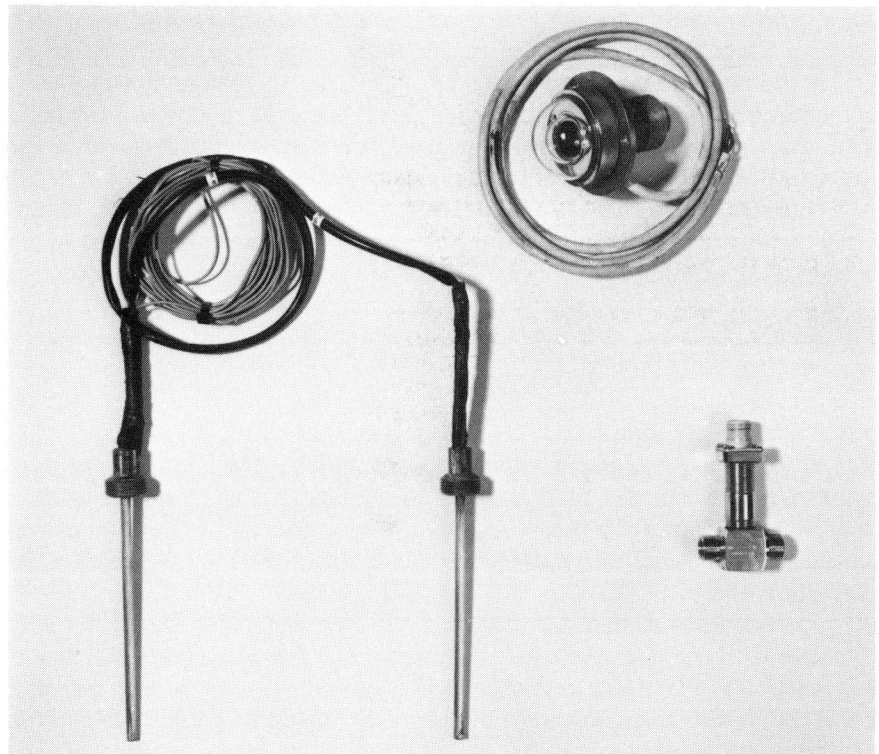

Figure 2.2. Three popular types of transducers used for solar-collector testing. Left to right — 4 + 4 junction thermopile, solarimeter and turbine flow meter

corresponding to a particular radiation intensity. If a flat-plate collector is tested under a range of irradiance conditions, scatter may be expected to be greatest in the low-efficiency region: the abscissa intercept, although conceptually significant in that it determines the stagnation temperature of a collector under any given conditions is not a particularly useful parameter in practice.

These difficulties have until recently been of little concern because measurement accuracy has been such that except for collectors that exhibit a markedly variable loss coefficient any curvature or separation of the lines has been swamped both by random scatter of data points and by systematic errors between different laboratories. The fact that most test methods have stipulated minimum irradiance levels has also helped here but relaxation of this restraint has, in at least one case, highlighted the care needed in interpreting test data. A couple of years ago a UK laboratory obtained results covering the whole efficiency range and found, much to its surprise, that the data points fell on a perfect straight line. It was, however, pointed out that since the full range of $\Delta T/I$ had been traversed with the collector never having risen more than a few tens of degrees above ambient the results gave no indication of its possible behaviour at higher absolute temperatures. This is illustrated in Figs 2.3 and 2.4 which are taken from a 1975 internal BRE report. Since these early days instrumentation accuracy has been a recurring problem and the conclusions of a 1977 BRE paper [16] need little modification in the light of experience. In addition to the systems outlined, there is now an increasing acceptance of platinum resistance thermometers for temperature difference measurement, although in air collector testing distributed thermopiles are likely to be preferred.

Turbine meters are still favoured for flow measurement and some groups install them vertically to minimize problems with entrained air. Interest in the completely different approach of using metering pumps is growing, but does not yet represent a tried and tested technique in the UK.

Determination of solar irradiance to better than 5% has proved troublesome not only because of calibration changes of individual instruments but as a result of differences between various calibration centres. These problems are now much better understood than five years ago but there remains the difficulty that currently available instruments cannot really be trusted to better than a couple of per cent even when used by experts. It seems unlikely that this situation will improve over the next few years so the only option for most laboratories will remain that of using currently available hardware and taking care in its calibration and use.

In Europe attempts to use ASHRAE or NBS test methods [14, 15] met with some success during the exceptional summers that followed the oil-crisis years. It was realized, however, that such conditions were unlikely to become an enduring feature of the Northern European climate and so attempts were made to develop and prove test methods in which the requirement for strong steady sunshine could be relaxed. Amongst these were a method for determining the heat-loss characteristic indoors and using sunlight to determine the intercept efficiency

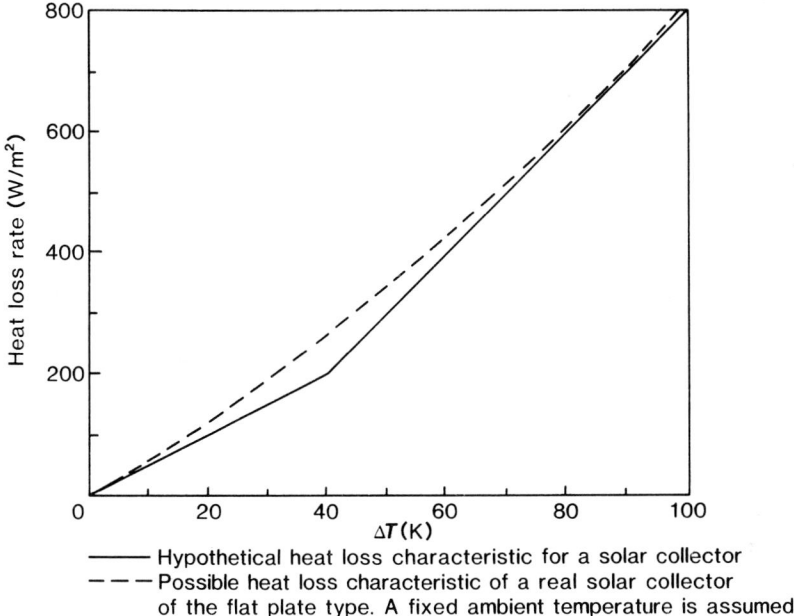

Heat loss rate (W/m²)

ΔT(K)

―――― Hypothetical heat loss characteristic for a solar collector
― ― ― Possible heat loss characteristic of a real solar collector
of the flat plate type. A fixed ambient temperature is assumed

Figure 2.3. Solar collector heat loss

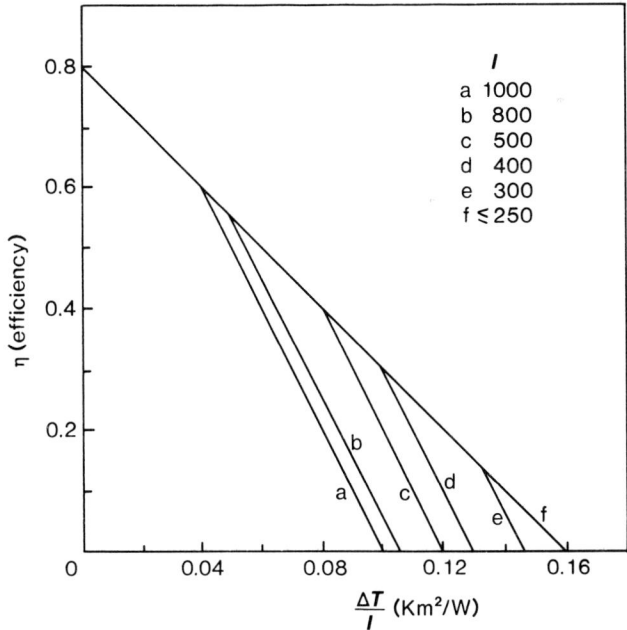

η (efficiency)

	I
a	1000
b	800
c	500
d	400
e	300
f ≤ 250

$\frac{\Delta T}{I}$ (Km²/W)

Figure 2.4. Efficiency curves derived from the hypothetical heat loss characteristic of Fig. 2.3, plotted for various values of the solar intensity I. η_0 *is assumed to be 0.8*

only. Since 1975 a collaborative group working under the auspices of the EEC has pooled its working experience and as a result a consensus document covering some of the problems and possible solutions of collector testing in Europe was published in 1980 [17].

In the UK work has centred not only around the EEC programme but on the development of new test methods under the auspices of the British Standards Institution. The principal effort has been devoted to a highly mathematical approach to analysing thermal performance data obtained under variable weather conditions. In this method the collector is rarely if ever at a steady state and each component is either gaining or losing heat at a rate determined not only by the immediate irradiance level but also by past events. Early work, some of it carried out at BRE, demonstrated that the response of a typical flat plate collector under such conditions could not be modelled using a single 'lumped' value of heat capacity and a single time constant but that the distributed nature of the capacity would have to be considered. Rather than force the collected data to obey preconceived ideas the method can generate, given a sufficiently wide range of internally independent input data, a transfer function that represents the behaviour of the collector. Of necessity this function also embodies the response

Figure 2.5. The outdoor solar test facility at BRE. The rotatable platform in the foreground is used to help assess collector testing methods. The roof-mounted collectors form part of simulated domestic hot-water preheat systems

factors of the instrumentation – especially that of the temperature difference and irradiance transducers – but this is a minor problem. If sufficient data is collected a linear collector characteristic can be produced and early experimental work has shown that the method has considerable promise. Difficulties do arise, however, because the method necessarily allows testing under conditions in which a sizeable fraction of the irradiance is diffuse. This contrasts with the ASHRAE approach and for results from these different methods to be compatible corrections have to be applied to account for the different transmission properties of cover materials for direct and diffuse radiation. For the experimentalist there is the problem of measuring direct and diffuse radiation on sloping surfaces with a range of incidence angles, and for the theoretician the need to convert these measurements to an equivalent normal irradiance, that is, the normal irradiance that is calculated to have the same effect on the thermal behaviour of the collector as does the received irradiance.

The method can be extended to encompass variable heat-loss coefficients but given that most outdoor solar collector testing is not yet at a stage where consistent linear fits can be obtained it might be considered premature to worry about fitting to higher orders, especially in view of the small errors involved for many flat-plate collectors. It is partly in the area of reproducibility of results that another approach to collector testing – that of using a solar simulator or artificial sun – is proving its worth. The only large solar simulator in the UK is operated by University College Cardiff and is described elsewhere [18].

For the purpose of this brief review it is sufficient to note that although the simulator approach has its own idiosyncrasies their effect on the data can be removed mathematically, and with some confidence. It is increasingly apparent that the repeatability achievable with this technique is so improved when compared with much outdoor work that second-order equations become a viable feature of the data processing.

On the practical side it should not be forgotten that the principal *raison d'être* of collector efficiency curves is that computer simulations can predict how different collectors will perform in any given system. Production tolerances, especially between batches of collectors having selective absorber surfaces may limit the usefulness of results obtained on any given unit, and since it can be entirely reasonable to expect degradation of thermal performance with age in some designs meaningful assessment will only be possible when (and if) robust methods of determining collector durability become available. This problem is further discussed in a later section. In assessing the future of solar collector testing several aspects stand out as particularly significant, and it should also be noted that this whole approach has not found universal acceptance [19].

(a) One of the most promising methods of outdoor testing for the UK appears to be that based on mathematical processing of data collected under variable weather conditions.

(b) The problems of instrumentation accuracy are such that commercial laboratories undertaking collector testing will have to work to a high standard for credibility to be assured. Testing is therefore unlikely to be cheap, especially in view of the limited home market. If calibration and operating procedures are allowed to become relaxed there could easily be a return to the days when random and systematic errors dominated most attempts to differentiate between nominally similar collectors.

(c) It may have to be accepted that for outdoor work the reproducibility attainable may be insufficient to differentiate reliably between similar collectors. This problem is not ameliorated by the possibility of continuing and perhaps variable systematic errors between different laboratories.

(d) The continuing refinement of methods using state-of-the-art solar simulators may render outdoor testing redundant for high-accuracy work. Development of inexpensive simulators able to give reproducible results without using sophisticated equipment may find favour with collector manufacturers. Exceptions may, for a time, arise where the collector's response to diffuse radiation is important or not easily calculable.

SIMPLE MONITORING OF INSTALLED SYSTEMS

One of the other areas of interest for performance monitoring is that of simple instrumentation that can tell a householder whether a solar water-heating system is working correctly, or at least whether it has developed any major fault. Lack of space precludes a full discussion but Table 2.1 shows what are thought to be the most common failures to affect thermal performance together with the types of simple instrumentation that could be used for detection purposes.

It is worth noting that control system faults can be difficult to detect principally because there is no easy way of checking the calibration of the collector sensor. The fact that a fault may progressively become worse over a period of years can also inhibit correct diagnosis. Details of the necessary safety and performance requirements for control systems have been published by BRE [9].

At the present time it seems that the most useful simple instruments that can be incorporated are:

(a) A variable area flow meter (pumped systems only).
(b) An hour-run meter (to record pump run hours).
(c) An impulse counter (to record pump starts).
(d) Warning lights on the control box.
(e) Dial thermometers.

RELIABILITY AND DURABILITY: THE WAY AHEAD

Most proponents of solar heating technology agree that active systems used under UK conditions will have long payback periods. For present purposes it is

Table 2.1. Simple instrumentation for fault checking or indication in solar systems. × = suitable indication; ? = possible indication.

Indication	Fault						Comment
	Partial air locking	Low flow	No flow	Fuse	Control system fault	Low system pressure	
Lights or buzzers			×	×	?	×	Could be part of automatic fault-detection system
Flow rate indicator			×				Put the money towards a proper flow meter
Flow rate meter	?	×	×				Equipment not available for thermosiphon systems
Flow switch		?	×				Could be part of automatic fault-detection system
Storage temperature	?				?		Requires interpretation
Pump run-hours		?	?		?		Requires interpretation
Pump starts		?	?		×		Can warn of incorrect control system settings. Has proved useful in practice
Pressure gauge						×	Only appropriate for system types
Silicon cell			?		?		Could be part of automatic fault-detection system
Comment	May be undetectable except by panel inspection	Caused almost exclusively by poor design and/or installation			Difficult to check. Fit and forget systems required	Very easily checked	

sufficient to note that for successful mass-market application lifetimes exceeding ten years may have to become credible. It is not difficult to postulate detailed designs of systems that, apart from minor maintenance of surface coatings and replacement of wearing parts such as pump bearings and valve seals, should last decades.

Against this background, it is useful to assess what could be done to improve the reliability and durability of installed systems. The problem is twofold. All components should be suitable for their duty, and compatible both with one another and with adjacent building components. More difficult is the development of methods for proving that components and systems have these properties. Herein lies a problem the many solutions to which have remained largely in the realms of empiricism. It is notoriously difficult to have confidence that the result of some accelerated ageing tests bears any relation to what will happen in practice. For example, if it were known that a collector would be subject to thermal shock about once a day then, providing there were good grounds for supposing that subsequent to occurrence of a shock no relaxation mechanisms operated to gradually relieve any induced strains, it would be reasonable to cycle the collector 1000 or more times per day, thus enabling ten years life to be simulated in less than a week. The catch is whether it can be known reliably whether there is a true dead-time in which nothing of significance occurs.

Tolerance to chemical or radiation exposure is an even more difficult area. Acceleration of reactions may be attempted either by increasing reagent concentration or temperature or both. For simple reactions it can be predicted what will happen but for materials of interest under real conditions of use the problem is largely intractable to meaningful theoretical resolution. Increase in the intensity of exposure may induce breakdown simply because repair mechanisms that are adequate under normal conditions become swamped, or completely different reactions can be induced.

A further and quite distinct difficulty is that although tests for real-time effects such as water penetration can be tailored to predict a known result on one component they can yield spurious results on a different but similar component. There exists an anecdote of a series of rain penetration tests for roof tiles that produced leakage through a design that experience had shown was suitable for areas of high driving rain.

Another specific example is provided by the known failure in service of collector plates manufactured from thin aluminium. A BRE experiment designed to assess the long-term behaviour of this design has now been running at high temperature for over three years. Lack of any failure to date may illustrate the wide range in durability of systems that are susceptible to cleanliness of installation.

Work in this whole difficult area started in earnest in the UK solar community following general agreement that the reliability and durability of collectors and systems was likely to prove at least as important for the development of the industry as was thermal performance. Tests can be divided broadly into two categories. First, the component is subjected to every extreme condition or combination of conditions that is likely to be met in practice. Such tests are sometimes called qualification tests and aim to ensure that no obvious mistakes have been made in selection of materials or fabrication methods. Had an agreed series of such tests been available during the early years of the UK solar industry some designs would never have reached production. Typical of the design faults that could have been detected are double-glazing units that crack when the collector is first exposed to strong sunlight, and absorber-plate surface coatings that cannot tolerate plate stagnation temperatures. Use of polystyrene insulation directly behind absorber plates also produced difficulties. Lest it be thought that this type of problem was unique to the UK an NBS study [20] provides evidence to the contrary, and also illustrates the American preference for large-scale testing. Secondly, and more contentiously, collectors can be subjected to accelerated ageing tests.

In the area of materials performance and general structural integrity it is likely that inspection by a competent and experienced organization will remain a useful if slightly arbitrary method of assessment. Such advice is not particularly useful to the development of standard test methods but there is a wealth of knowledge available on the behaviour of materials and a few standard tests can be a poor substitute for years of hard-won experience. For practising engineers sympathetic to this general approach the BRE solar book combined with the BRE Digest series, for example, provides a good grounding in both solar-related and general building problems.

FINANCIAL ASPECTS: SYSTEM COSTS AND SYSTEM DESIGN

There is a fundamental difference between financial and economic analyses. A financial analysis considers only the monetary aspects of a project as seen by one or more of the parties involved. An economic analysis may, depending upon its scope, encompass considerations of resource depletion, nuisance and option values and if all relevant factors are included should produce an optimum answer. Some of the popular misgivings about economic analysis as applied to energy conservation have recently been discussed [21].

In the early days of the UK solar industry much argument revolved around possible energy savings. Following publication of the British Standard there has been widespread acceptance of the potential limitations of solar domestic-water heating in the UK and the discussion has turned both to reliability and durability and to reduction in capital costs. By far the largest field trial of solar water-heating systems in the UK is being carried out by BSRIA under contract to BRE and forms part of the Department of Energy's solar heating programme. Costs were carefully recorded and are thought to be representative of 1979 building industry practice. Table 2.2 shows a breakdown of costs, on a per house basis, corrected so as to exclude charges attributable to monitoring equipment. Since

Table 2.2. Installed costs of 4.5 m² solar collector systems and possible cost reductions (1979 prices)

	Costs	Possible reductions
	£	£
Installation work	390	190
Solar collector	270	190
Steelwork for roof	82⎫	50
Lead flashing for roof	70⎭	
Copper cylinder (227 1)	67	50
Control system and sensors	39	20
Pump	20	20
Pipe	12	10
Incidentals	20	10
Total	970	540
Equivalent to	£215/m²	£120/m²

some of these figures relate to bulk purchases, usually in lots of forty or eighty, they probably represent an optimistic picture for a single house installation.

In the author's estimation the final figure of £970 for a system incorporating a 4.5 m² collector could be reduced to £540. A major part of this saving could come from development of streamlined installation techniques. Further reduction, say to £350 probably represents an unattainable target for an industry tied to UK labour rates and satisfactory application of current technology. At 1979 fuel prices the systems could be expected to save about £45 per year for users of on-peak electricity or perhaps £30 per year in a more average case. A 1979 cost of about £350 may therefore be needed for mass market application of this type of system unless there is a significant increase in the price of energy in real terms.

These field trials (Fig. 2.6), coupled with others in the BRE programme, have served to emphasize both the importance of good workmanship and the desirability of design rules not being too rigid. As an example of the latter topic the well-known rule of thumb of 50 l of preheat storage for every square metre of collector will be briefly discussed.

It seems obvious that for any solar water-heating application there is an optimum volume of water storage. Put simply, insufficient storage will lead to high temperatures but very low collection efficiencies and too much storage will produce only tepid water. Given a set of assumptions it is possible to derive an optimum value and there are several examples of this in the literature. Courtney [7] shows a value of about 50 l/m², Brinkworth [22] has discussed the topic in more detail and gives a range of 50–80 l/m² whilst the present author has postulated 35–50 l/m² as being an 'economic' optimum [23]. In practice, as with other optima derived from cosy discussions of idealized equations, there is only a slight chance of all the components being available in exactly the right sizes. In

Figure 2.6. Three of the houses at Basingstoke fitted with 4.5 m² flat-plate collectors for domestic hot-water supply

discussions of insulation the argument then broadens to consider the desirability of overspecifying components especially those that would be difficult to upgrade at a later date [24] but solar water-heating systems need a slightly different approach. Given the continuing modification of solar collector design it is unlikely that after a few years life a system could be extended using the same type of collectors. The option would therefore be one of having a completely new set of collectors or of accepting aesthetic and possibly plumbing difficulties. Providing that durability can be assured this argument favours installation of larger rather than smaller systems if it is thought that fuel prices will increase and is in direct conflict with other UK assessments. Storage volumes may be increased retrospectively either by substitution or addition, and in some cases twin stores have been used as original equipment in restricted spaces [25].

In many cases, however, available space will limit storage to below the technical 'optimum' of 50 l/m^2. This factor has emerged both from field trials in existing houses and from industry contributions to BSI discussions. Installation of additional water storage in roof spaces carries the penalty of a deadleg between warm and hot stores and experience indicates that attention to details such as insulation and structural integrity can be less good if components are located in relatively inaccessible areas. The requirements for support of additional loads in trussed rafter roofs are known but perhaps not always observed and proper standards of work will necessitate appropriate capital outlay.

The BS Code does not contain detailed information on the penalties to accrue from sub-optimum storage: area ratios, but some guidance can be obtained from both references [7] and [22]. The observation that 'with a given hot water demand the (system) performance is only weakly dependent upon the storage capacity, once this has exceeded a fraction of about three-quarters of the mean daily demand' [26] is worth noting. When considering systems intended to last ten or twenty years or longer it is advisable to contemplate whether the use to which they may be subjected could change over that time. As already suggested a possible scenario embraces higher fuel prices and reduced hot-water usage as two aspects of an increased awareness of the need to conserve resources. If this were to occur the optimum design of solar water-heating systems could change and although it would be unwise to estimate precisely the direction this might take (if only because of uncertainties in component and fuel costs) one possibility is that typical domestic systems could have more collector area and less storage than at present. Several qualitative arguments can be used here. Water storage costs, and to some extent installation costs, are representative of industries that are already well established. There is a considerable difference between the cost of similarly sized polypropylene cisterns and copper cylinders but both have been used for hot-water storage. Solar collectors are largely still the product of small companies and within Europe there is little mass production. If the solar market increases there is, therefore, scope for cost reduction in collectors and, more especially, for higher performance: cost ratios. This is further discussed in the next section. Some American estimates show collector and system costs reducing

Figure 2.7. Close-up of a collector mounting point of the type shown in Fig. 2.6. The strength of the steelwork together with the large area of lead flashing make this system one of the most expensive yet used in the UK

considerably from present levels but in view of the range of subsidies and tax credits that are available some care is necessary in translating this experience. A recent Canadian review [27] shows 1979 collector system costs between $450/m² and $600/m² with projected reductions down to $300/m². Calculation of these reductions assumed measures that might not be feasible for UK solar water heating, and the figures provide a sobering adjunct to some of the recent BRE optimism.

The disadvantage of low storage:area ratios, apart from possible sub-optimal performance, is that excessively high temperatures can be produced more readily in systems not incorporating suitable safeguards. Over-temperature protection for a range of system types is discussed in the BRE book and one of the neatest solutions is perhaps the use of combined cylinders in which the effective storage:area ratio increases when high temperatures are attained. Such systems can also ameliorate the performance penalty to some extent. Another approach would be to use a combined water and latent-heat store having a transition temperature between 50 and 60°C. Such a system could limit peak temperatures without wasting energy available on very sunny days but would have to be both durable and very cheap for mass-market application.

THE UNCERTAIN FUTURE

In a recent book an economist argued that the only thing to be learnt from energy forecasts is that while all are likely to be wrong, some will be farther off the mark than others [28]. This is not a new observation, but it is worth repeating.

The major problem facing the proponents of renewable energy technologies is that there is no certainty that they will be required for centuries. Similar uncertainty is inherent in predicting the requirements for conventional energy supply as was recently emphasized in the architectural press [29]. What seems increasingly certain, however, is that present levels of energy consumption can be reduced significantly without affecting the prosperity of a developed country. Clearly, the scope for such action depends on how wasteful are present habits and some possible developments in the UK have been discussed elsewhere [30].

It has been suggested that the prospect of a large overseas market for solar collector technology, especially in less-developed countries should give rise to sponsored development of the industry in the UK. This is a reasonable line of argument in respect of countries able to pay for these goods and services. The uncertainty here is clearly considerable and may not be greatly reduced by improved technical knowledge.

A sobering observation is that over the last few years the prices of solar water-heating systems in the UK have more or less kept pace with fuel prices. For example in 1977 with a domestic on-peak electricity cost of 2.5 p/kW h BRE quoted £700–£800 as a typical capital outlay for a well-engineered, 5-m^2 system whilst in 1980 the corresponding figures were 3–4 p/kW h and £900–£1500 with higher prices not unknown. Such comparisons must always be made with care because of the distorting influence of taxes, subsidies and marketing policies but there may be a strong underlying link between the cost of fuel and the cost of both manufactured goods and labour.

It is against this general background that the future for the solar collector industry in the UK must be assessed. To achieve penetration of the UK mass market on any significant scale using only current technology might require a reduction in the installed costs of good systems of perhaps a factor of 2 to 4 together with an improvement in the general standard of installation work. It does not appear that these goals will be attained in the near future and the UK market will probably remain dominated both by sales of swimming pool systems and by sales of domestic preheating systems to those who have surplus capital to invest and who wish to minimize their personal consumption of conventional energy. The former market has an obvious saturation limit and the latter is probably small. Future developments could of course enhance the prospects and it is theoretically possible for sophisticated non-tracking collectors to deliver up to twice the energy of typical flat-plate designs of equal size when used in domestic water-heating systems. If such collectors could be manufactured using automated techniques and a minimum of expensive raw material it is therefore possible for a doubling of output to occur for no increase in total system cost.

Experience to date has indicated, however, that collectors able to achieve such a performance are likely to be expensive even in mass production and the uncertainty of a real volume market has been a factor in determining whether manufacturing facilities will be constructed within Europe.

A likely course of development is therefore that sophisticated collectors will be manufactured in experimental quantities and sold in the up-market sector where cost effectiveness in respect of energy saving alone is not a criterion for purchase of prestige equipment.

The application of these systems should, however, not be allowed to overshadow continued use of much simpler collectors. Faced with an increasing choice of collector types, system designers will need credible and widely available computer programs (or authoritative nomograms derived from these programs) to help select the optimum type or types of collector for any given application.

The present situation, in which collector selection and sizing is much a matter of matching available financial resources to available roof areas or storage spaces, proved adequate enough for the initial growth phase of the solar industry, but cannot instill the confidence necessary for more widespread application of industrial and commercial systems.

An important proviso in the use of computer techniques is, however, that predictions can be precise to ten decimal places and accurate to none and the data now becoming available from accurately monitored field trials and laboratory systems will serve as vital corroborative information.

ACKNOWLEDGEMENT

The work described has been carried out as part of the research programme of the Building Research Establishment of the Department of the Environment and this paper is published by permission of the Director.

REFERENCES

1. Heywood, H. (1954) Solar energy for water and space heating. *Inst. Fuel. J.*, 334–52.
2. Heywood, H. (1971) Operating experiences with solar water heating. *JIHVE*, **39**, 63–9, 144.
3. Courtney, R. G. (1976) *An appraisal of solar water heating in the UK*, BRE Current Paper, CP7/76, Garston, UK.
4. Long, G. (1976) *Solar energy: its potential contribution within the United Kingdom*, Energy Paper 16, Department of Energy, HMSO, London.
5. Heating and Ventilating Contractors' Association (1979) *Solar Heating for Domestic Hot Water. Guide to Good Practice*, London.
6. British Standards Institution (1980) *Code of Practice for Solar Water Heating Systems for Domestic Hot Water*, BS 5918, London.
7. Courtney, R. G. (1977) *A computer study of solar water heating*, BRE Current Paper, CP30/77.

8. Sharpley, D. (1977) Testing of solar collectors and systems in development and production. *UK ISES Conference C11*, April 1977, London.
9. Wozniak, S. J. (1979) *Solar Heating Systems for the UK*, HMSO, London.
10. Freund, P. and Wozniak, S. J. (1980) BRE tests heat pump and solar at Basildon. *Building Services and Environmental Engineer*, 3(3), 26–9.
11. Wozniak, S. J. (1979) *Solar Heating Systems for the UK*, HMSO, London, p. 14.
12. Page, J. K. (1979) Solar energy – A personal view. *Sun at Work in Britain*, **No. 8**, Pergamon Press, Oxford, pp. 41–4.
13. Wozniak, S. J. (1979) *Solar Heating Systems for the UK*, HMSO, London, p. 78.
14. The American Society of Heating, Refrigerating and Air-Conditioning Engineers (1978) *Methods of Testing to Determine the Thermal Performance of Solar Collectors*. ASHRAE Standard 93–77 (corrected printing), New York.
15. National Bureau of Standards (1977) *Provision Flat Plate Solar Collector Testing Procedures*. Report NBSIR 77–1305, Washington, USA.
16. Wozniak, S. J. (1977) Measurement in solar collector testing. *UK-ISES Conference C11*, April 1977, London.
17. Commission of the European Communities (1980) *Recommendations for European Solar Collector Test Methods (Liquid Heating Collectors)*.
18. Gillett, W. B., Rawcliffe, R. W. and Green, A. A. (1980) Collector testing using solar simulators. *UK–ISES Conference C22*, London, pp. 57–71.
19. Shurcliff, W. A. (1979) *New Inventions in Low Cost Solar Heating*, Brick House Publishing, Andenes, Mass., USA, pp. 249–53.
20. National Bureau of Standards (1977) Solar energy systems – survey of materials performance. Report NBSIR 77–1314, NBS, Washington, USA.
21. Fisk, D. J. (1979) The economics of energy conservation in buildings. In *Buildings the Key to Energy Conservation*, (ed. G. Kasabov), RIBA Publications, London, pp. 12–13.
22. Brinkworth, B. J. (1978) Active collection and use of solar energy, in *Ambient Energy and Building Design*, pp. 51–9, Construction Press, Harlow.
23. Wozniak, S. J. (1979) *Solar Heating Systems for the UK*, HMSO, London, p. 60.
24. Fisk, D. J. (1976) *Energy conservation: energy costs and option value*. BRE Current Paper, CP57/76, Garston, UK.
25. Wozniak, S. J. (1979) *Solar Heating Systems for the UK*, HMSO, London, Plate 32.
26. Brinkworth, B. J. (1980) British Standards for solar heating, *UK–ISES Conference C22*, London, pp. 99–111.
27. Walker, H. V. and Macalik, M. J. (eds). (1979) *Energy Conservation Design Resource Handbook*, Royal Architectural Institute of Canada, Montreal, Section 3.10.
28. Bailey, R. (1977) *Energy – The Rude Awakening*, McGraw-Hill, London.
29. Shankland (1982) Energy impact. *Architects' Journal*, **171**(4), 180–7.
30. Wozniak, S. J. (1980) Use of renewable energy sources in the UK. In *Martin Centre Conference 'The Architecture of Energy'*, Construction Press, London.

DISCUSSION OF CHAPTER 2

Dr J. C. McVeigh (Brighton Polytechnic). It is possible to begin to detect from one or two manufacturers the reduction in the cost of solar collectors that Dr Wozniak hoped would occur from his 1979 figures, which indicated a cost then of £60/m². We will see on the market very shortly quality collectors with *selective surfaces* at about £60/m² (1981 prices). The stable price in a time of inflation indicates a positive reduction in cost. Is it the correct approach to look at the straight economic payback of the solar industry? Should some consideration not be given to the possibility of the solar industry creating jobs for school leavers in small decentralized industries? It could be argued rightly that this should be done for insulation first. It is the general approach that is being advocated as a real alternative to basing judgements on technical arguments alone.

Dr S. J. Wozniak (Building Research Establishment). The cost of solar collector systems to the consumer is certainly an important factor in assessing the possible future of the industry. Dr McVeigh quoted a possible future cost of £60/m². This would be for the collector alone. Labour costs for well-installed systems can be considerable, as has been confirmed in our field trials to date. Current total costs are between £250 and £350/m² but even if there was to be a reduction to £120/m² at 1979 prices there does not seem to be the prospect of a very significant market in the UK in the near future unless there is a marked increase in the price of fuel in real terms.

Dr R. W. Todd (National Centre for Alternative Technology). Dr Wozniak said that solar collectors could only provide low-grade energy which was not in itself suitable for the manufacturing processes necessary to make further collectors when the originals fell apart in the future. This is valid but it only becomes significant when all high-grade energy sources are needed for high-grade purposes. Until that point the possibility always exists of switching high-grade sources displaced by solar heat to high-grade uses. Even in the distant future the probability exists that renewable high-grade sources will be available, and provided solar systems can yield many times the energy required for their manufacture they can still be important contributors to our energy needs.

I am slightly concerned with some of the comparisons being made between solar heating systems and heat-pump systems. Clearly in a practical sense they are alternative choices, but from a resources point of view they are very different. Solar heating systems provide a complete alternative to fossil fuels and non-renewable sources. At the moment, heat pumps have co-efficients of performance less than three. If these are electrically driven, as most of those in use are, using electricity produced from fossil fuel with an efficiency of 30% they result in approximately the same fossil fuel use as if it were burnt directly for heating.

Dr S. J. Wozniak. Certainly at the present time much high-grade energy is used to provide low-grade heat. However, this does not alter the fact that flat-plate collectors could not be self-renewing in an energy sense in the long-term future.

Looking many years ahead, it is possible that buildings might become increasingly dependent upon electricity if only because fossil fuels might be reserved for applications such as transport. Within a scenario embracing far greater use of nuclear power, electrically driven heat pumps might become widely used in buildings.

R. Cullen (Cullen Carter and Hill). In connection with the application of heat pumps in housing and the problem of balancing the National Grid System, is there any future in storage of off-peak electricity using a heat pump and water storage which distributes the heat during normal tariff hours, and therefore should improve the peak-load demand position?

Dr S. J. Wozniak. Domestic space heating is not responsible for the one and only peak on the electricity grid in this country. There are essentially two difficulties; the first is the seasonal imbalance in electricity usage, and the second is the occurrence of sharp peaks at certain hours of the day. It would not be difficult to prevent heating systems contributing to the latter, but if the seasonal imbalance of domestic electricity usage became a very serious problem then it could be worthwhile changing tariffs so as to penalize consumers who used electricity during winter months. If this were to occur, then as with long-term thermal storage we might have to re-think a lot of our ideas as to what constitutes an optimum system.

I agree that a heat pump might be cost-effective in this country where the alternative is oil or direct use of electricity. It is doubtful whether at the present time it could be cost-effective where the alternative is mains gas. The concept of using existing off-peak tariffs to drive a heat pump to charge a thermal store is a concept that is incorporated into one of the BRE experimental low-energy houses.* Whether such a system would be cost-effective in any given situation is not an easy question to answer. It would probably depend upon the building type and also the particular use of the building.

* See also Mountford, D. J. and Freund, P. (1982) The performance of a well insulated house heated by an air–water heat pump. *International Journal of Ambient Energy*, **3**, pp. 9–18.

3

Use of Ambient Energy in European Climates

H. Hörster

INTRODUCTION

The climate in Europe, especially in Middle Europe, is characterized by moderate temperatures during summer and winter, by a low solar insolation in winter and a moderate insolation during summer. The use of ambient energy in buildings is therefore different compared to sunny regions in Europe and even colder but sunny regions in the northern part of America.

This chapter deals with measures to reduce the energy demand in buildings by improved insulation, use of solar energy by passive and active means (window systems, solar heating and hot water systems) and by heat pump systems in different operational modes.

WEATHER

Figure 3.1 gives an overview of relevant weather data for four locations from north to south in Europe: Stockholm, Hamburg and Freiburg (West Germany) and Carpentras (France, near Marseille). The figures show the distribution of the average of the monthly insolation on a south-facing surface with an inclination equivalent to the latitude of the location. The yearly insolation on this surface and the heating degree days (based on 19°/15°C) are indicated in the upper part of the figures. The number on the right-hand side is the ratio between the insolation of the three best summer months and the three worst winter months. The weather in Hamburg is similar to the weather in most parts of Great Britain.

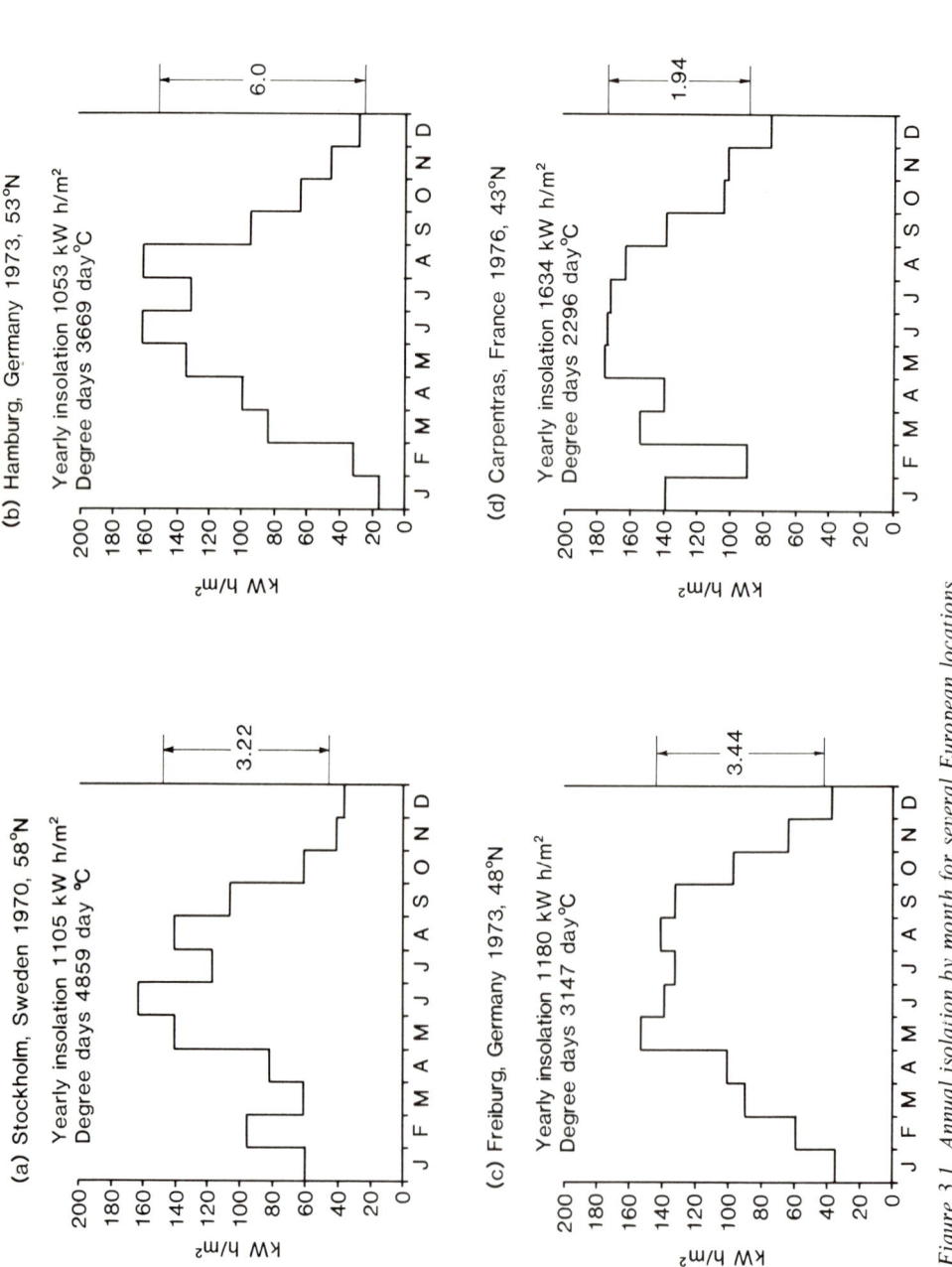

Figure 3.1. Annual isolation by month for several European locations

HEATING DEMAND OF BUILDINGS

The performances of heating systems, which make use of ambient energy, depend strongly on the specific heating demand of a building. The demand is influenced by the weather, the user habits and more important by the building standards. In most European countries the improvement of building standards is under way or under discussion. On the other hand buildings constructed according to older standards will exist for a longer period. Therefore the performance of ambient energy systems will be discussed for three building standards:

(a) 'Normal' representing existing older buildings,
(b) 'Swedish' according to the 1975 Swedish standard,
(c) 'Experimental' according to the Philips Experimental House in Aachen, West Germany, which comes close to the limits achievable.

Figure 3.2(a) shows the thermal parameters of the three standards including the air ventilation rates. These parameters are used to compute the heating demand of an average single family house (floor area $= 160$ m^2) which is considered to be a model house. The three standards are applied to this house. Figure 3.2(b) shows the monthly heating requirement. The results of the yearly demand (upper part of the figure) demonstrate the possibilities of energy savings in buildings by applying adequate insulation standards. Internal load of about 25 kW h per day (American standard) plus solar radiation through windows have been taken into account.

USE OF SOLAR ENERGY IN BUILDINGS

Passive use

The passive use of solar energy in buildings – that is the reduction of the heating requirement by utilizing solar energy by larger south-facing window areas – is also being discussed in Europe. The passive use has the advantage of using solar energy at all intensity levels, but the disadvantage of difficulties in the control of the thermal comfort inside the house. Buffer zones, shutters or large heat capacities are needed to handle the indoor climate. Little information is known about the energy-saving potential of different window/shutter combinations in a south-facing wall.

Figure 3.3 gives the results of a computer simulation of different window qualities on the yearly heating requirement in Hamburg 1973. Parameters are the three different building standards; starting point is the building without any windows.

First, different window systems are installed in the north-, east- and west-facing walls. After this, the effect of the south-facing window is shown as function of the window area. The results show that for the Normal house all window/shutter

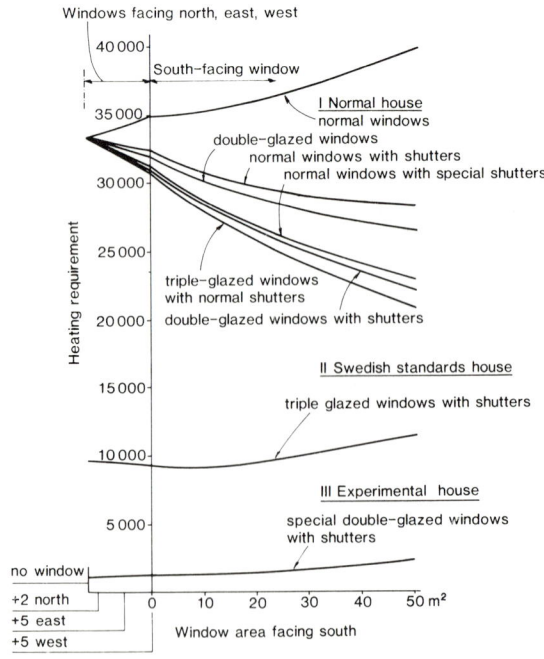

House type	Normal	Swedish	Experi-mental
Heat transfer coef (W/m² K)			
Walls	1.12	0.37	0.17
Roof	0.5	0.26	0.18
Windows (day/night)	5.8/5.8	2.2/1.5	1.9/1.2
Number of hourly air changes	1.0	0.5	0.3 (eff.)
Transmissivity of window	0.9	0.74	0.7
Internal heat capacity (kW h/K)	4.0	2.4	7.0

Floor area: 160 m²
House volume: 400 m³
Window area: 26 m²

34995	Normal house
9091	Swedish standards house
1301	Experimental house

(a) (b)

Figure 3.2.(a) Definition of standard houses and (b) monthly heating requirement in Hamburg (1973)

Figure 3.3. Heating requirement as a function of the window area (Hamburg, 1973)

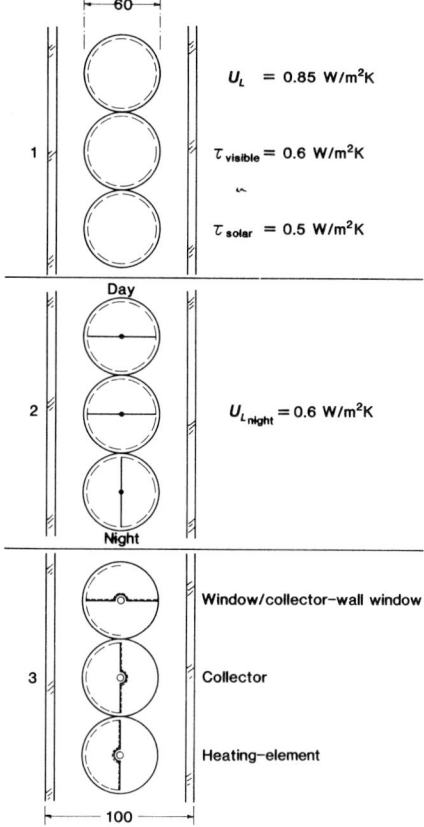

Figure 3.4. Transparent-wall consisting of a row of evacuated tubes with (1) heat reflecting mirrors on the inner side of the tubes; (2) a tunable shutter; (3) a heat-transfer pipe acting like an active/passive heating element

combinations give energy savings except the single pane (normal) window. For the building with the Swedish standard only window systems better than triple-glazed windows with shutter give savings. The explanation for this is the reduced heating season (see Fig. 3.2) for the better insulated buildings which is caused by the reduced heating requirement in combination with the internal load.

An interesting transparent-wall concept has been investigated by the Philips Research Laboratory in Aachen. It consists of a row of evacuated tubes with heat-reflecting mirrors on the inner side of the tubes. The tubes are located within the space of a double-pane window with outer dimensions of about 10 cm (Fig. 3.4). The transparency for visible light is 60%, for solar radiation about 50%. The heat-loss coefficient has been measured to 0.85 W/m² K. In an improved version a tunable shutter is incorporated within the tubes, allowing the

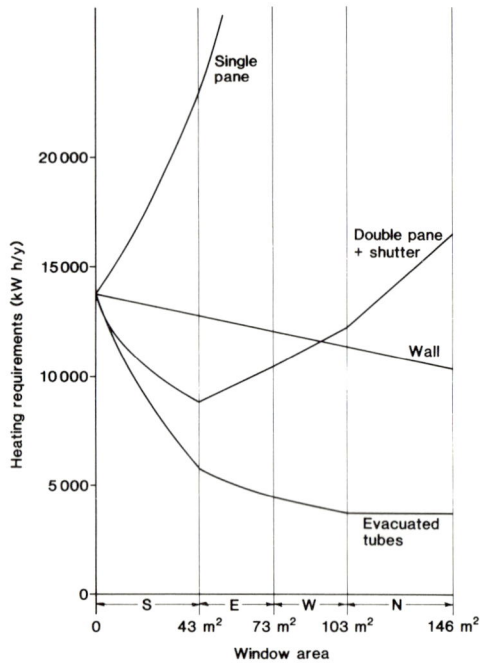

Figure 3.5. Yearly heating requirement as a function of window area and orientation for successive replacement of walls by windows in a SH-house in Stockholm 1970 (50% IEA-load)

control of the solar gain and the indoor climate automatically (Fig. 3.4, part 2). With a metallic coating on the shutter surface (e.g. aluminium) the U_L-value in the closed mode of the transparent wall can be reduced to $U_L < 0.6$ W/m² K. This is similar to a wall of 25 cm thickness, consisting of porous cinder concrete block. The concept can be further improved by incorporating a heat-transfer pipe through the evacuated tubes (see Fig. 3.4, part 3). By special arrangements of the heat mirror and a non-selective absorber on one side and a metallic reflector on the other side of the tunable shutter this wall acts as a solar active as well as a passive wall and a heating system.

To demonstrate the advantage of the transparent-wall system (with a tunable shutter) the effect on the yearly heating requirement of a single family house in Stockholm (Sweden) is demonstrated in Fig. 3.5. The x-axis indicates the glass surface of vertical walls starting from south-orientation to east, west and north. The figure indicates, therefore, the heating requirement of a house with no windows (starting point for $x=0$) and a house with all vertical walls (146 m²), made from glass with a thermal quality as indicated. The line 'wall' gives the results, obtained by improving the wall of the house with no windows from $U_L = 0.3$ to 0.2 W/m² K. The results demonstrate the possibility of window systems with adequate optical and thermal properties as an important energy-saving element even in the climate of middle Europe.

Because of the low thermal losses and the automatic control of the solar transmission this wall concept offers new possibilities to building designers and architects.

Active use

The active use of solar energy is discussed for a solar heating and hot water system for Freiburg 1973 (southern part of Germany). The performance of a solar system depends strongly on the performance of the solar collectors used and the solar system, i.e. the arrangement of the buffer tank and the auxiliary heating system.

Figure 3.6 shows the solar system used for simulation. It is an air-heating system which operates on the lowest possible heating temperatures resulting in

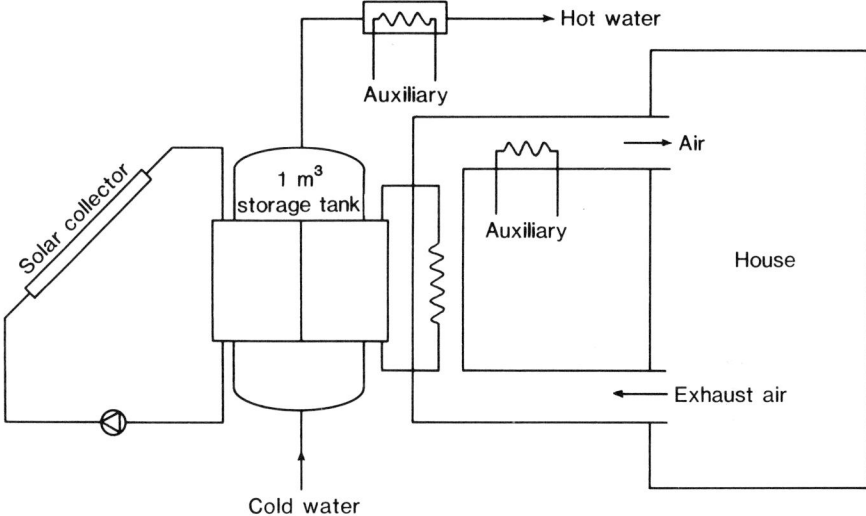

Figure 3.6. Solar system

highest system efficiencies. The energy demand for the hot-water production is 4000 kW h per year for a family of four. The collectors used for simulation are a standard collector (single pane, non-selective) and a high-efficiency collector (tubular, evacuated, selective). These collectors cover the field of non-focusing, low-temperature collectors.

Figure 3.7 gives the result of the percentage of the solar energy for heating and hot water as function of the collector area. Parameters are the three building standards and the collector quality. For a maximum collector area of 50 m² the solar percentage for the Normal house with standard collectors is only 25%. This figure seems to be too low to interest a home owner in this solar application. The

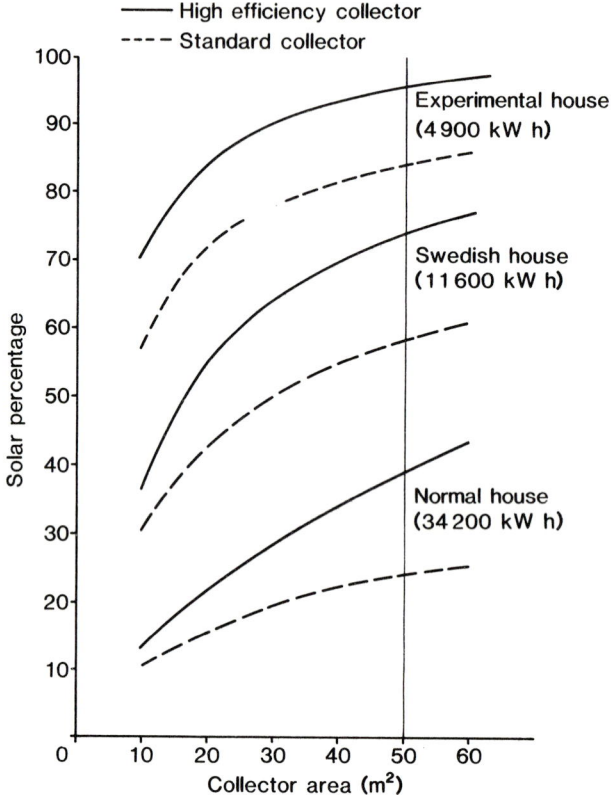

Figure 3.7. Percentage of solar energy for heating and hot water. Storage volume 1 m³; weather data – Freiburg, 1973

solar percentage is increased to 38% by using high-efficiency collectors. Higher percentages can be obtained for the Swedish house and the Experimental house.

In Figure 3.8 the system's efficiency is shown as function of the collector area for high-efficiency collectors. The system's efficiency is the ratio of the solar energy used and the solar insolation on the collector surface. Assuming system's costs of 500 DM/m² per collector area (which are optimistic future-orientated costs) a solar energy price for a pay-back-time of about twelve to fifteen years can be calculated. The results are shown on the right-hand side of Fig. 3.8.

The Normal house has the highest system efficiency and the lowest solar energy price, but according to Fig. 3.7 an insufficient solar percentage. On the other hand the Experimental house standard gives low systems efficiencies and consequently very high solar prices. The results indicate that there will be an optimum with respect to solar percentage and solar energy price for a house with a standard between Normal and Swedish.

Figure 3.9 shows the result of such an optimum solution where for an improved

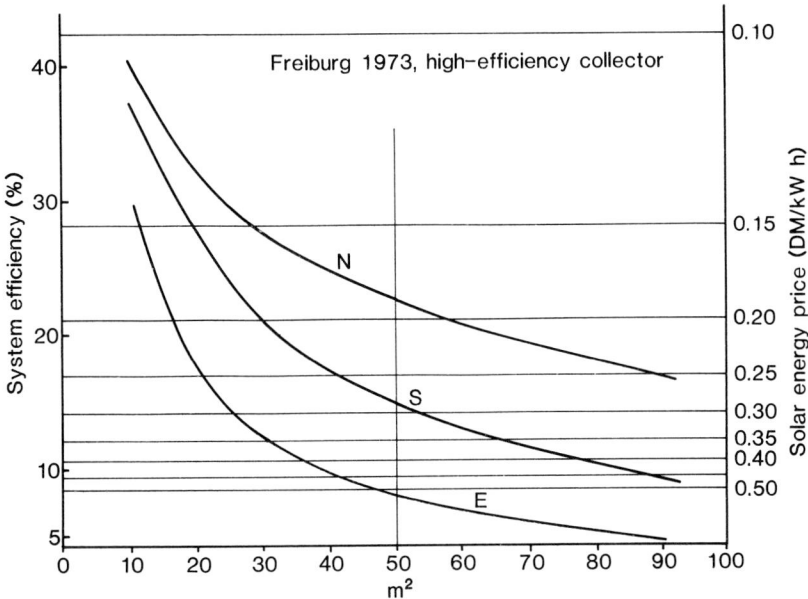

Figure 3.8. System efficiency and solar energy price (Freiburg, 1973). Storage volume 1 m³; system cost 500 DM/m²

Normal house with an energy demand of 20 300 kW h, a solar percentage of 50% and a solar price of 0.18 DM kW h for a collector area of 38 m² high efficiency collectors can be obtained.

HEAT-PUMP HEATING SYSTEMS

Heat-pump heating systems which make use of ambient energy are in use in Northern and Middle Europe and in large parts of the US. Most of the systems are air–liquid or air–air systems using the ambient air as the thermal-energy source. Recent solar-assisted, heat-pump systems are under discussion in Germany. In these systems a non-glazed absorber is the evaporator of the heat pump which collects different forms of energy, e.g. solar energy, convective and condensation energy.

A wide use of electrically driven, heat-pump heating systems in a monovalent operation may cause peak-power problems for the utilities. Therefore, bivalent heat-pump systems are common in Germany where the heat pump operates above 3°C and the heating system is switched to an oil furnace below 3°C. These systems have a high COP and a relatively small heat-pump capacity. They are used in existing houses where the oil furnace is already present to save oil and heating costs.

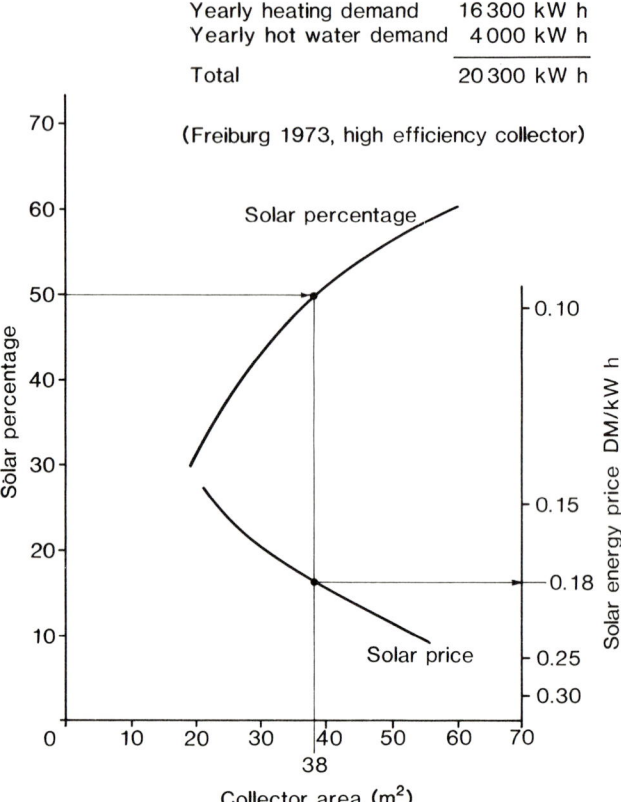

Yearly heating demand	16 300 kW h
Yearly hot water demand	4 000 kW h
Total	20 300 kW h

(Freiburg 1973, high efficiency collector)

Figure 3.9. Solar heating and hot water application. N-house improved standard

The performance of a heat-pump system is described by the following parameters.

(a) Percentage of the heating requirement delivered by the heat pump. This figure depends for a bivalent system on the balance point (switching point from heat pump to the oil furnace), on the building standard and the weather.

(b) Seasonal performance figure (SPF) represents the ratio of the thermal energy delivered to the house heating system and the electric energy used by the heat pump and the fan. This figure also includes energy to defrost, tank losses and switching losses of the heat-pump system.

Figure 3.10 shows the air–liquid system which is used for simulation. The buffer tank (~ 1 m^3) reduces the on/off cycles and hence switching losses of the heat pump. Tank temperature is variable depending on the outside temperature.

Figure 3.11 gives the results of this system (weather data: Hamburg 1973), in the upper part the SPF as function of the balance point and in the lower part the percentage of the heating requirement of the three different house standards. The

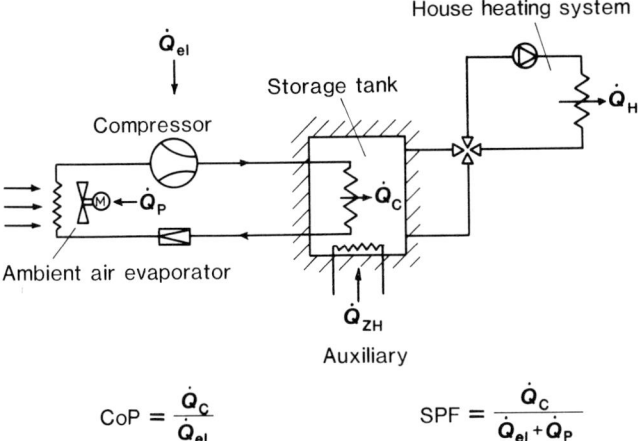

$$\text{CoP} = \frac{\dot{Q}_C}{\dot{Q}_{el}} \qquad\qquad \text{SPF} = \frac{\dot{Q}_C}{\dot{Q}_{el} + \dot{Q}_P}$$

Figure 3.10. Air–heat-pump heating system used for simulation

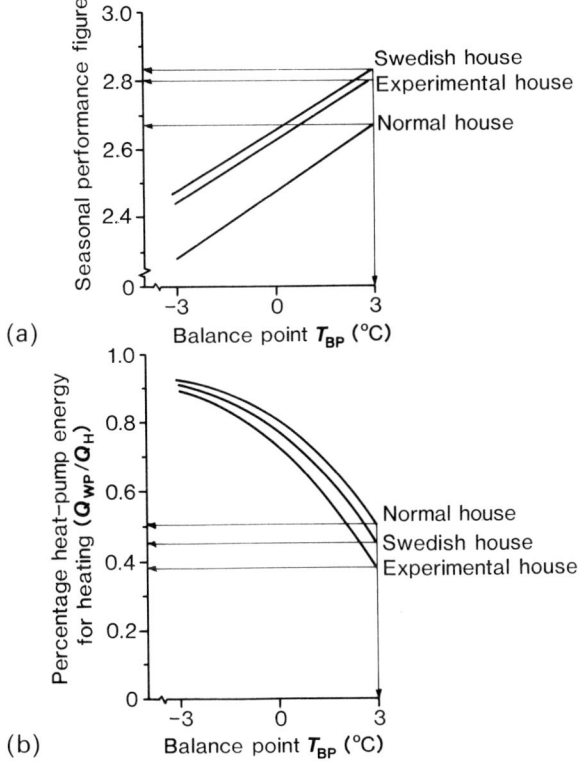

Figure 3.11. Ambient-air–heat-pump system. (a) Seasonal performance figure (SPF); (b) percentage heat pump energy for heating requirement

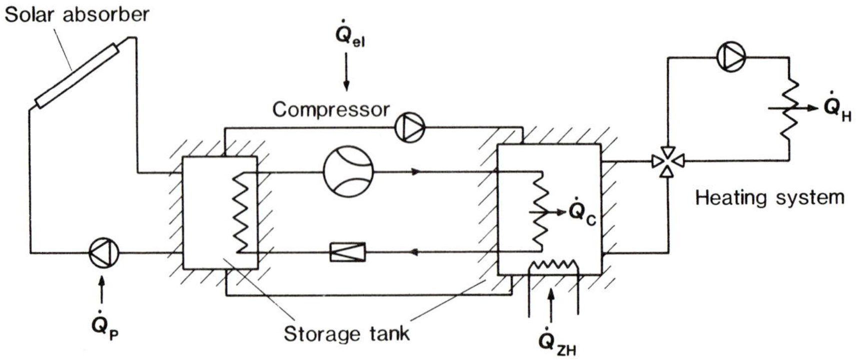

Figure 3.12. Heat-pump heating system with absorber

results indicate the strong influence of the balance point on these heat pump parameters within the temperature range from $+3°C$ to $-3°C$.

Figure 3.12 shows the heat-pump system with an absorber where an additional buffer-tank between absorber and evaporator is installed.

Figure 3.13 gives the results of the analysis for the Swedish house, in the upper part the SPF as function of the absorber area, in the lower part the yearly energy flows. The SPF is nearly the same compared to the air heat pump system and shows a weak dependency on the absorber area. According to the lower figure, solar energy is the main energy source for the absorber. Only for small absorber areas ($\sim 10\ m^2$) is convection as important as solar energy. Direct solar gain and condensation-energy from water vapour in air are negligible. From these results it can be concluded that there are only a few arguments for a solar-assisted, heat-pump system compared to an air system, which is more compact and less expensive.

CONCLUSIONS

The use of ambient energy in buildings to reduce the energy requirement for space heating and hot-water demand has to start with the improvement of the building standard. For the example of a single family house it has been demonstrated that the application of the Swedish Building Standard results in a reduction of the heating demand by more than 60% compared to Normal houses. The residual energy requirement can be cut by half by solar systems or heat pump systems.

FURTHER READING

Bruno, R. (1978) Which models for what? *Proc. Austr.-German Workshop on Solar Energy Systems Design, ISES* (German section), Dusseldorf.

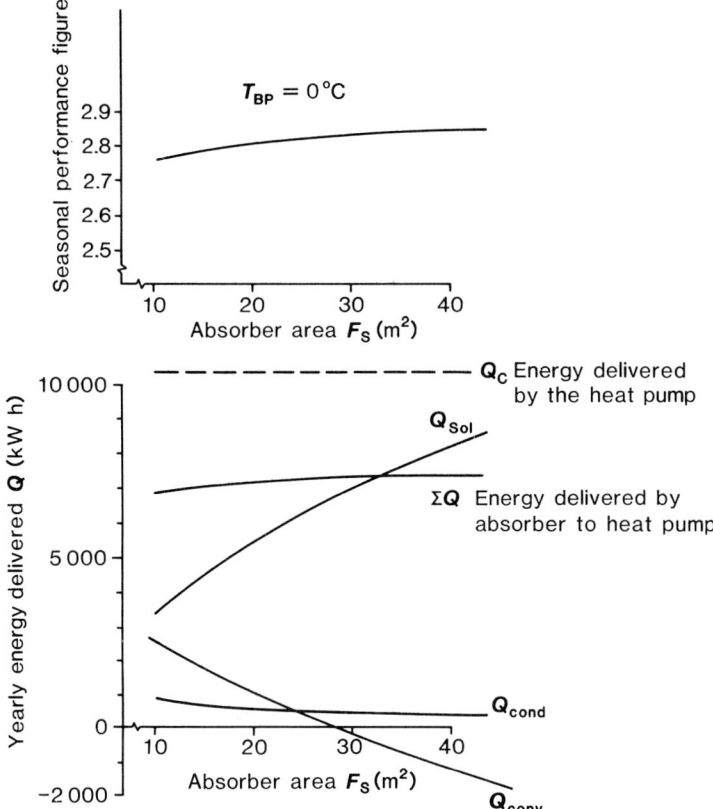

Figure 3.13. Heat-pump system with absorber. S-house, Hamburg, 1973

Bruno, R. and Steinmuller, B. (1979) On calculation heating and cooling requirements, *Energy and Buildings*, **2**, 197–202.

Hörster, H. (1980) *Ways to an Energy Saving Building*, Philips Fachbuchverlag, Hamburg (in German).

Hörster, H. and Kersten, R. (1981) *Proc. of the Conference on Recent Developments in Solar Collector Design*, *ISES* (UK section), London.

Knabben, H. (1980) Performance of heat pump systems. *Proc. ISES* (German section), Aachen (in German).

DISCUSSION OF CHAPTER 3

Dr J. C. McVeigh (Brighton Polytechnic). Dr Hörster's discussion of Fig. 3.13 differentiates between solar energy – the main energy source – and direct solar gain (negligible). An explanation of the exact meaning of the two terms would be helpful.

Dr H. Hörster (Philips GmbH Forschungslaboratorium, Aachen). 'Solar gain' is the total solar energy transmitted through the system via the heat pump, the direct solar gain is the amount of solar energy which can be used *directly* without upgrading by the heat pump.

J. Keable (HELIX Multiprofessional Services). Both Dr Hörster and Dr Wozniak (Chapter 2) mentioned the problem that as building performance improves and in consequence energy demand diminishes the demand is at a time when solar energy receipts are lowest. This is a particular concern to the electricity supply industry in connection with the use of heat pumps but it is also a problem for solar collector manufacturers. Will the current research on various ways of interseasonal storage of solar energy be more likely or not to solve this problem than using alternative fuels for heat-pump operation such as oil or gas engines to avoid the increased requirement for electricity at the coldest time of the year.

Dr H. Hörster. Peak load is a vital element in the economics of electricity production. The generating capacity required must have a certain annual running time to produce cheap electricity.

In using solar energy for active space heating, normal houses have the disadvantage that space-heating demand is so high that as a consequence the solar percentage is low. The contribution of solar energy is increased by improving these existing houses and an active solar system may then give good results if solar collectors providing high efficiency at a low cost are available. Well-designed and well-insulated houses like our experimental house are quite different. For these houses the use of solar energy for space heating makes no sense, therefore active solar systems should be used just for hot-water supply. In such 'zero energy' houses electricity can be used in very small heat-pump systems which because of the efficiency of operation and the consequent very low power consumption would avoid peak power. Such a future scenario can be discussed but it would involve changing all our houses. There would be no need for extra power stations. Nevertheless it is possible to believe in this future. I guess in the next ten years there will be a tremendous market for heat pumps and solar systems without peak load problems if buildings are first improved to reduce the energy demand for space heating to 50–60% of present requirement. Heat pumps or solar-active systems would then be very attractive options.

Dr S. J. Wozniak (Building Research Establishment). The development of cheap and effective long-term heat storage is certainly a major problem for some solar-energy systems. However, its successful development could radically alter the prospects for other technologies also. The problem of long-term storage is highlighted very well in Dr Hörster's chapter. In the experimental house the 50 m² of collector are able to work at only about 8% efficiency over the year. If long-term storage could be incorporated the house could probably be completely heated using 15 or 20 m² of collector.

Dr H. Hörster. If seasonal storage were available the position would be changed tremendously, but there is no real commercial solution in prospect. All the storage devices under investigation in research establishments around the world are costly and complicated. One interesting solution under discussion in Sweden is to use a very large storage volume by grouping a number of houses around a large storage tank. Cost-effectiveness is also a problem with this possibility but we believe solar energy could be used more widely by investigating applications other than solar-energy systems in single houses.

Another Swedish project that looks promising is to use solar energy from a large collector field connected directly with a district heating system without any storage. The solar contribution is only 10–15% of the total energy demand of this district heating system but if it pays it is the right application. The application where solar energy makes sense from a commercial viewpoint is the one to look for. If solar energy pays, the solar industry can expand, collectors come down in price with mass production and other possibilities become commercially viable, for example, the use of solar energy in houses for hot water and heating systems will look much better.

Dr R. Jackson (Institute of Energy). An example that appealed to me some years ago was the suggestion that if you decided to save money by turning off your electric chandelier for an hour and use candles instead, then the match needed to light the candles would cost more than the electricity saved – not necessarily an accurate statistic then or now, but it does illustrate the importance of realism.

Dr K. W. Garrett (Redland Technology Ltd). Energy roof/heat-pump systems are selling well in Germany at the moment but as Dr Hörster said the average COP is not very different from an air-source heat pump. However, a major point in their favour is that they have a water-source heat pump so there is very little noise and vibration associated with tnem. Hence the system can run at night and take advantage of cheap-rate electricity.

Dr H. Hörster. The question is what *is* the acceptable noise level of a ventilator. If someone is very sensitive to noise the price difference of about £2000 might be worthwhile to them, but there is no advantage from the absorbing roof in performance.

P. F. Brown (Isle of Wight Area Health Authority). The monthly heating requirements shown in Fig. 3.2 for the three types of house in Hamburg are normally pro rata apart from August when the experimental house appears to require some heating. Is this, in fact, energy required for cooling because the house was getting too hot?

Dr H. Hörster. Sometimes heating in summer may be needed although it has little influence on annual requirements. If the thermostat is set for $20°C$ and the house has a certain heat capacity, a period of cold days in August may be such that there is a requirement for heating.

Dr B. Smith (Brunel University). Were the heat transmittance factors indicated for the vacuum tube walls experimental or theoretical? Perhaps Dr Hörster would also comment on the economics of such a system.

Dr H. Hörster. The values were experimental, a full set of possible configurations were measured and I believe the figures are conservative. Cost effectiveness is not always the best concept. We prefer the question in reverse: if a saving is possible what total capital investment will it support? The vacuum tube wall has extra advantages beyond heat saving. The thermal transmittance/wall thickness ratio (U/L) is low compared to a concrete wall (the wall thickness just 10 cm), which would be a plus point to architect and builder, the surface temperature is close to the air temperature – good for thermal comfort, as a consequence of the small U/L the heating system can be readily calibrated – it can also be smaller, therefore cheaper. If everything is taken into account, the transparent wall may be a cost effective possibility, compared to normal windows it looks quite an interesting alternative even from a price point of view.

Dr J. Twidell (University of Strathclyde). In Fig. 3.1 of Dr Hörster's paper the insolation/m² has been plotted. Such data is often given for a horizontal surface, which is the normal meteorological way of plotting insolation. Is there not always a case to consider insolation on vertical surfaces as more important for buildings, or indeed direct insolation?

I would also suggest that the monthly heating requirements vary considerably with latitude. For example, in Britain there are very big differences between southern England and northern Scotland. It is important, therefore, to realize that data in the form of Fig. 3.3 and 3.7 will have a latitude dependence.

Dr H. Hörster. I agree fully with Dr Twidell. In presenting results the basis is always given. The weather data used for Hamburg is real weather data on an hourly basis for 1973. The solar radiation on the roof is on a south-facing surface. The tilt angle corresponds to the latitude, 58° for Stockholm for example and 53° for Hamburg.

Dr S. J. Wozniak (Building Research Establishment). I emphasized in Chapter 2 that there may be no one optimum design of passive solar building for the UK. Over the small range of latitude of the UK there is not a great variation in insolation for the year as a whole but in winter there is a factor of 3 or 4 between the south of England and the north of Scotland. It seems to have been agreed that in a well-insulated house the heating season could be reduced essentially to those months in which there is little solar radiation. It is not immediately obvious whether solar space heating will first be viable in the north or south of the country but it may be dependent upon the use of the building.

4

How to Finance Ambient Energy Systems in Local Authority Projects

Anthony Kirk and Andrew Burke

OVERALL STRUCTURE

Whilst the local authority operates within the structure of the public sector of the UK economy, the strategy and tactics to be considered in financing ambient energy schemes should not be unduly different from those that would be adopted by a large public company operating in the private sector.

Local authorities are numerous – 457 in England and Wales split up into 39 Metropolitan Counties (such as Durham, Nottinghamshire and Surrey), 6 Metropolitan Counties (such as Greater Manchester and West Yorkshire), Greater London with the 32 Boroughs and the City of London, and 8 Welsh Counties (such as Dyfed and West Glamorgan. In addition are the District Councils for non-Metropolitan and Metropolitan Councils such as Tunbridge Wells and Leeds.

A complicated picture providing a variety of services, including education, housing, social services, libraries, sports facilities and refuse collection. The common denominator to these diverse services is *buildings* and buildings use energy.

It is important to define the scale of finance, as local authorities are large property and vehicle fleet owners spending considerable sums on building and, increasingly, in re-habilitating old buildings.

'Capital works' can be taken as one definition of this financial scale which is generally split up into acquiring land and existing buildings, new construction, vehicles, plant and other items. Capital expenditure for the year 1978/79 shows that the Counties, the Metropolitan Counties, the Welsh Counties and Greater London (excluding the London Boroughs) spent £57.8 million on land and existing buildings, and £508.5 million on new construction.

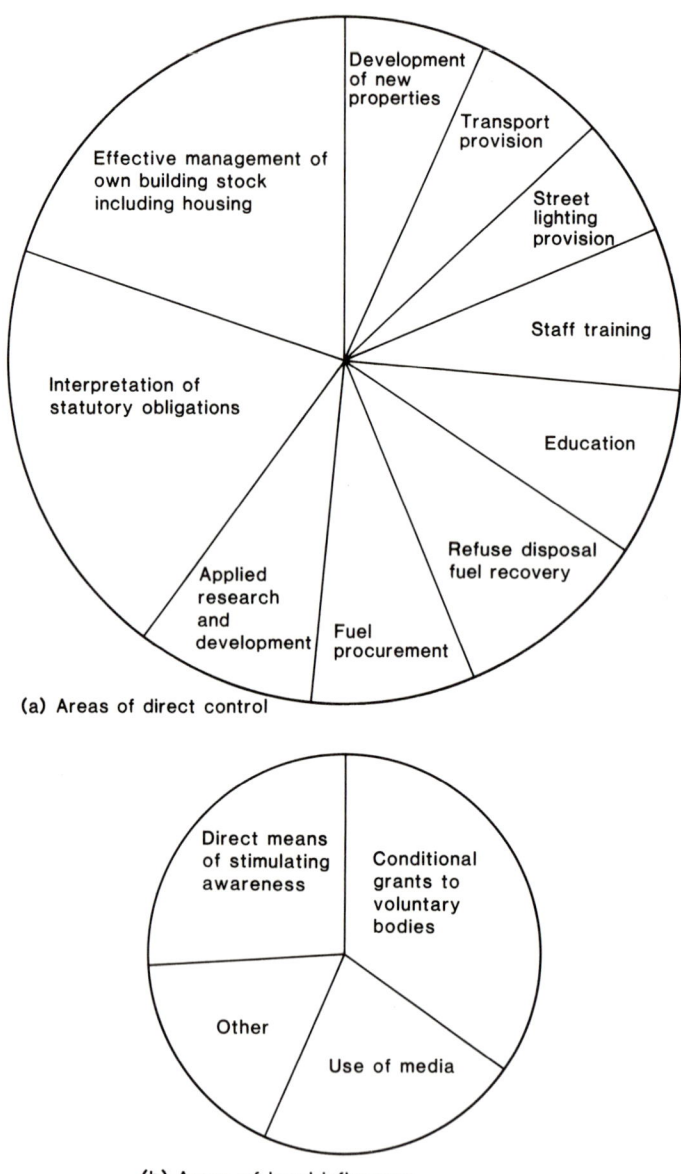

Figure 4.1. Local authority energy policy – scope for action. Based on the Standing Technological Conference of European Local Authorities, UK Energy Conservation Interest Group Guidelines for a Positive Local Authority Energy Policy

Despite the cut backs, local authority expenditure is considerable and the opportunities for energy savings within their areas of influence are numerous. Figure 4.1 shows the scope for action and is taken from the paper: *Guidelines for a*

(c) Mutual organizational influences

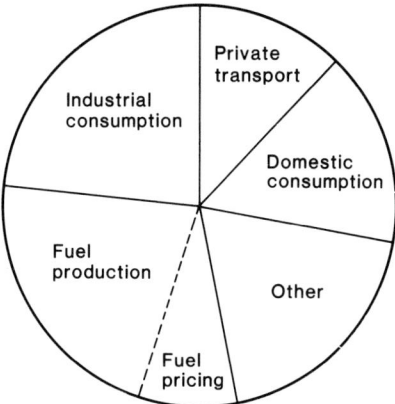

(d) Areas of limited influence

positive local authority energy policy, Fig. 4.1(a) deals with the areas of direct control, (b) areas of local influences, (c) mutual organization influences, and (d) areas of limited influence.

It cannot be said that local authorities are without the opportunity of carrying out new ambient energy schemes. But how can they be financed within the collective public sector cash limits or individual project's budget?

First, before the project is considered it is essential that the Chief Executive and his Board of Officers have an Energy Policy that is clearly defined, and supported by each of the disciplines. It is equally important that the elected representatives from the leader of the Council to each Committee Chairman understand the important part they play from a motivation point of view.

Secondly, it is generally agreed that the hardest part of any project is establishing it, and for this early period the commitment of Senior personnel is essential; once 'the will to work' is established, the ways and means naturally follow.

Thirdly, how is this commitment nurtured and encouraged? Experience has shown that one way found to be successful is for key elected representatives and officers to visit projects and see for themselves that particular energy ideas actually work. The commitment from the Secretary of State for energy towards solar heating is unlikely to be less now that he has run the solar tap on a project and felt the temperature of the water. The creation of public interest, providing it is responsible, cannot be over-emphasized and provides welcome support in taking a project through the financial approval stages.

Fourthly, it is also important the local authorities demonstrate practical, safe, well-designed and well-supervised projects that are examples to other sectors of the economy. This does not necessarily mean that new, exciting ideas cannot be developed; it means that they have to be well worked out, and done properly.

Fifthly, whilst the new, exciting ideas are always attractive, it is often the less exciting that are the more rewarding. The league table for energy saving would not be headed by either active solar collector systems, or wind or wave energy ideas at the moment.

Sixthly, and possibly the most difficult, is to do with training, education and re-education. Do we spend adequate time understanding the options available to us? Is it true that frequently the apparent simple options require a good deal of understanding, and the more complex ones require little understanding but often considerable maintenance. The examples of the thermosyphon principle, understood so well by plumbers for so long, has become nearly a lost art, whilst the pump and two-way valve are fixed *ad nauseum*. The transfer of heat through a building structure is complicated, yet the concept of passive solar gain in a building is simple and rewarding. It matters little who takes the lead; naturally as an architect I am biased towards that discipline, not on any professional or financial ground, but particularly because the design of good buildings is about architecture and there must be one individual who has overall responsibility to the client body.

Finally, how can ambient energy prospects be financed? is an artificial question and would better be re-phrased as: 'How can buildings which include various ambient energy measures be financed within the normal budget limits?' The success of energy-saving measures must be seen in the light of being within the overall cost limits and perhaps less on whether the individual element is cost effective.

The fourth part of this chapter (p. 80) suggests ways and means of analysing both the elements of the building from a cost point of view, and presenting them in a graphic form which can be easily understood, whilst applying a similar technique to energy analysis based on a modified calculation method.

However, before looking at the techniques of calculation and analysis and hard practical building advice, it is perhaps worth seeing what has been built over the last few years in the local authority field, including those projects in the pipeline.

EXAMPLES OF SOME AMBIENT ENERGY SYSTEMS IN THE LOCAL AUTHORITY SECTOR

The examples have been chosen to cover as wide a range as possible and are in eight categories:

(a) *Wind power:* the proposed Kingston-upon-Hull wind-powered housing project designed by the alternative technology group at the Hull School of Architecture.
(b) *Geothermal energy:* the proposed City of Southampton geothermal heating scheme in the City Centre development.
(c) *Solar domestic water heating* (DWH): the London Borough of Southwark's solar domestic water-heating trial in fourteen re-habilitated mid-Victorian

houses, and the earlier prototype at Penrith Close, London Borough of Wandsworth, both monitored by the South London Consortium.

(d) *Higher insulation standards:* the Kirklees Metropolitan Council's highly insulated new houses at Taylor Street, Batley, and the London Borough of Lewisham's Lawrie Park Road re-habilitation project which is part of the SLC Demonstration project funded by the EEC.

(e) *Solar DWH and passive solar gain:* the London Borough of Lewisham's projects at Brownhill Road with flats for old people designed by private architect Royston Summers, and the timber-framed new houses under construction for the London Borough of Southwark at Wingfield Street.

(f) *Passive solar:* the Giffard Park housing scheme for single people in Milton Keynes designed by Energy Conscious Design for the Society of Co-operative Dwellings to maximize the benefits from passive solar gain.

(g) *Heat pump:* the City of Salford and University of Salford low-energy housing project at Strawberry Hill incorporating a heat pump.

(h) *Education buildings:* a number of local authorities have made interesting and useful contributions in low-energy building design. Hampshire County Architects' Department has worked closely with The Martin Centre for Architectural and Urban Studies at the University of Cambridge and two schools were scheduled to be built in 1981.

Wind power

The Alternative Technology Group at the School of Architecture at the Hull College of Higher Education have designed a housing scheme for Kingston-upon-Hull City Council on a site at Bransholme for thirty-two dwellings which incorporates a number of interesting and novel ambient-energy ideas.

Hull is well placed to take advantage of wind energy and has an average wind speed of 5.2 m/s. The housing project incorporates high standards of insulation, doors and windows thus reducing air infiltration, and provides a heat recovery system for one-third of the waste water.

The estimated heat requirement for the houses is 3300 kW h/annum for space heating and 2400 kW h/annum for hot water. The aerogenerator output is distributed and controlled electrically to each house and there is stored for immediate or subsequent use in a thermal brick store 600 mm × 600 mm × 1.8 m high, 99 kW h capacity, incorporating a warm-air fan and control unit.

The warm-air output can be easily mixed with the output from the low-output coal burner (maximum output 4 kW – slumber at 0.8 kW) to provide sufficient back-up space heating when there is insufficient stored wind power. The domestic hot-water storage tank is fed from the waste-water heat recovery unit.

This project is particularly interesting in that it harnesses: wind energy; heat recovery; low-output, continuous-burning bituminous coal burner; high standards of insulation and air infiltration; and considers the problems as a group scheme.

Geothermal energy

The most favourable area in the UK for geothermal energy is the Hampshire Basin and years of survey work culminated in late 1979 when the Department of Energy, through the agency of the Energy Technology Support Unit, identified a site in the centre of the City of Southampton.

The City Council were about to develop the new Western Esplanade scheme on 50 acres of vacant land, reclaimed during the 1920s when the Eastern Docks were constructed. The new scheme which consists of a new shopping centre, bus station, extensive leisure facilities, commercial and hotel complexes and a new industrial area will be the first geothermal project in the UK.

The pilot hole to verify the existence of the hot water and test loading was carried out in 1979 on a site by the Marchwood Power Station and it was confirmed in March 1979 that water was found at a depth of about 1.5 miles. It had a temperature of 70°C and the water table brought it to within 500 feet of the earth's surface so that pumping will be required to bring it to the top.

A new production well will have to be drilled on the site of Southampton's new development, and because of the heavily saline composition of the water, and for other technical reasons, a second hole will be drilled to re-inject the 'used' water back into the earth, thus avoiding pollution of the nearby River Test and Southampton Water. Because of the aggressive nature of the water, it could not be used in any conventional heating installation and so, on reaching the surface of the earth this heat will be extracted by passing it through a titanium heat exchanger before returning it immediately by the re-injection hole. The temperature of 70°C requires special design considerations which will govern the manner in which the heat is distributed and radiated throughout the development of the scheme, and work is now in hand on the development of heaters to do just this job.

The life of the supply is estimated to be between thirty and fifty years at the end of which time all that is required is the drilling of a new production hole and the use of the first production hole for re-injection.

Solar domestic water heating

The Whateley Road, London Borough of Southwark domestic solar water trial (Figs 4.2–4.5) is funded by the Department of Energy and monitored by the South London Consortium. It was one of the first major local authority achievements in active solar energy in existing buildings and is based on the earlier research carried out at Wandsworth (Fig. 4.6).

The mid-Victorian houses are three storeys high including the semi-basement floor. The rehabilitation included removal of the back extensions, new drainage, wiring, replacement of lath-and-plaster ceilings and wall linings with plaster and plasterboard, and replacement of the roof slates – thus affording an ideal opportunity for integrating the solar panels in the roof.

Figure 4.3. Existing timber roof members exposed

Figure 4.2. Whateley Road before rehabilitation

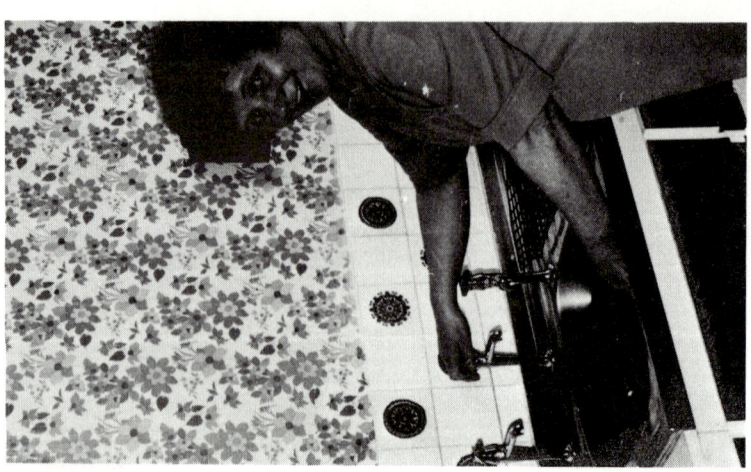

Figure 4.5. The renovated house showing how solar collectors are integrated into the roofscape

Figure 4.4. One of the tenants using the three-tap system

Figure 4.6. Askill Drive, Wandsworth, site of the first SLC solar domestic hot-water trial

The houses face directly south making them suitable for installing solar panels on the front roof. These houses are not rehabilitated to 'low-energy' standards. The only variation from the normal standards of rehabilitation in the district is that all the hot water pipes are lagged (Figs 4.7 and 4.8).

The layout of the solar DHW system is very simple. The solar panel is connected indirectly to the hot water cylinder situated in the basement airing cupboard next to the bathroom. The kitchen is situated above the bathroom. The hot-water pipes from the top of the cylinder to the services are approximately 3 m in length. The heat supplied by the solar panel is supplemented by an electric immersion heater located in the top half of the same cylinder. One of the houses has a two-cylinder, three-tap system.

Figure 4.5 shows how the scheme has been successfully integrated into the street pattern. Monitoring is being carried out for a further period under the guidance of the Energy Technology Support Unit ETSU at Harwell.

Higher insulation standards

Yorkshire Development Group and Kirklees Metropolitan Council
The Yorkshire Development Group (YDG), now the Yorkshire Liaison Group, and Kirklees Metropolitan Council have co-operated together to design and build a pair of five-person, three-bedroom, semi-detached houses on a new local authority development at Batley, West Yorkshire. The houses completed in May 1980 are Nos 10 and 12 North King Street and are part of a sixty-six house development and have incorporated at the design stage a number of energy-saving measures. The main purpose of the project is to show that higher standards of insulation reduce energy demand.

The two highly insulated houses are to have their energy requirements compared with two standard YDG houses (of which 15 000 have been built) constructed to normal building regulation insulation standards. The highly insulated houses were internally redesigned within the original shell dimensions to incorporate draught lobbies to both front and rear entrances, and to locate the living areas adjacent to the internal party walls, with the lobbies, staircases and utility areas adjacent to the external gable walls.

The housing project was financed through the Department of Environment normal housing cost yardstick negotiations and included the following measures in the highly insulated houses.

(a) Ground floor: 50 mm polystyrene sheets laid on concrete slab.
(b) External walls: 100 mm cavity (wider than usual) filled with polystyrene beads.
(c) Window jambs and head: 'Plasticell' expanded PVC cavity closers.
(d) Windows: sealed double glazing units incorporating integral single glazed vertical condensation strip and splayed reveals to ground floor lounge windows, compensating for smaller than usual size.

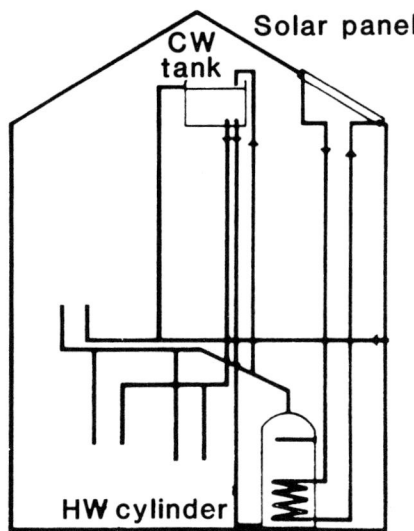

Figure 4.7. Domestic hot-water solar system in three-storey and mid-Victorian rehabilitated houses at Whateley Road, Southwark

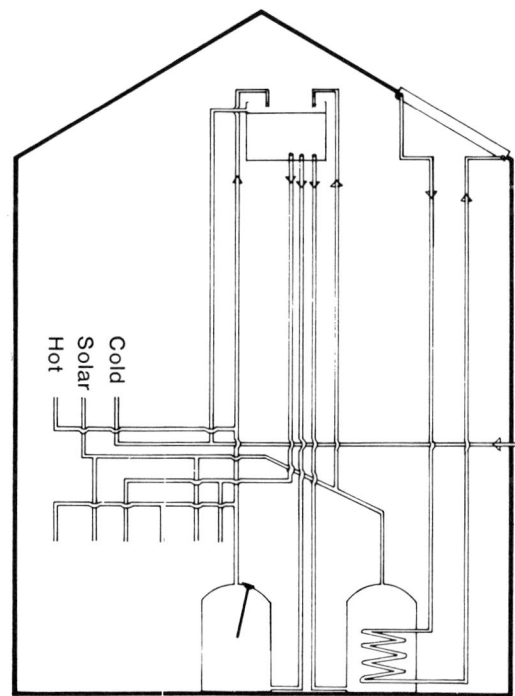

Figure 4.8. Solar layout for three-tap system

Figure 4.9. Two five-person three-bedroom semi-detached houses at Batley, West Yorkshire, showing condensation strip of single glazing in sealed double-glazed window unit. Ground floor window has splayed reveal to improve light penetration

(e) Roof space: 200 mm glass-fibre insulation quilting.
(f) In addition: windows reduced in size, external doors include solid flush ply with polyurethane core. All external doors and windows weatherstripped with schelgal foam. External and internal lobby doors filled, with automatic door closers.
(g) Heating for low-energy gas house: Johnson and Starley gas warm-air system, with a Janus III gas water heater.

Heating for low-energy electric house: electric storage heaters and convector heaters using off peak 24-h tariff electricity.

Heating for standard R 5.3 YDG house: heating and hot water are provided by gas-fired, low-pressure, hot-water, central heating. A wall-mounted boiler, three elves radiators and an indirect cylinder.

The demonstration project is being monitored for two years by the School of Architecture at Leeds Polytechnic and will include the humidity and ventilation losses as well as a record of the air temperatures at selected levels.

London Borough of Lewisham, Lawrie Park Road

The low-energy projects at Lawrie Park Road, London Borough of Lewisham illustrate the opportunities that are available for new forms of funding, and both schemes are part of the EEC and UK Department of Energy Demonstration Projects.

The project is aimed at bringing together known energy-saving measures to design, build, monitor and demonstrate Low-Energy Houses as Integrated Systems within the UK Government cost limits.

The two low-energy developments, one (Fig. 4.10) the rehabilitation of three Victorian houses being converted into fifteen flats, and the other (Fig. 4.11) eighteen new houses, are now completed. The design studies indicate that a theoretical energy saving of about 70% for space heating can be achieved within the current cost limits, if an integrated approach to the house design is applied. The main reason for this is that by rearranging various elements of the houses there is often a 'trade off' between the cost of one element set against the saving of another. Under certain circumstances, it is possible to offset part of the cost of energy saving measures against the saving made in the capital cost of the reduced heating system, thus keeping within the cost limits.

Analysis (Fig. 4.12) has shown that nearly as much energy is lost through ventilation and draughts for most standard houses as through all the other elements put together. This element therefore needs special attention. By providing draught lobbies for the two outside doors and by draught stripping all windows and doors, the loss through ventilation and draughts is reduced by 50%. Also, the warm air heating system costs considerably less than the 'wet' central heating system usually installed. This measure, therefore, not only saves energy but also saves on capital cost.

Another point made is that modest energy-saving measures evenly distributed over the house are more effective than one or two measures only implemented to excess, e.g. double glazing and little else. Evenly applied energy-saving measures, i.e. walls, underfloor and loft all moderately insulated and draughts reduced from all the windows and doors, plus a more rational heating system, eventually result in lower temperature gradients thus reducing the heat loss further by not allowing the highest temperatures nearest the coldest surfaces. Absence of cold radiation

Figure 4.10. Existing Victorian housing at Lawrie Park Road, Lewisham, which is to be rehabilitated into low-energy flats as part of an EEC energy demonstration project

and of draughts has another indirect effect in that comfort levels are reached at a lower temperature. Every degree C drop in temperature represents between 5% and 10% saving in fuel.

Various studies carried out by SLC Energy Group showed that by adopting an integrated approach to the design of low-energy houses, it was possible to build them within UK Government Cost Yardsticks. Details of the financial analysis carried out by an independent firm of Quantity Surveyors for the new houses in the project along with the details of both rehabilitation and new developments, energy analysis etc., are available for scrutiny.

Solar (DWH) and passive solar gain

London Borough of Lewisham, Brownhill Road

The recently completed housing scheme at Brownhill Road, South London, for the London Borough of Lewisham shows what can be done within normal local authority cost limits. It is an excellent demonstration of tenacity and

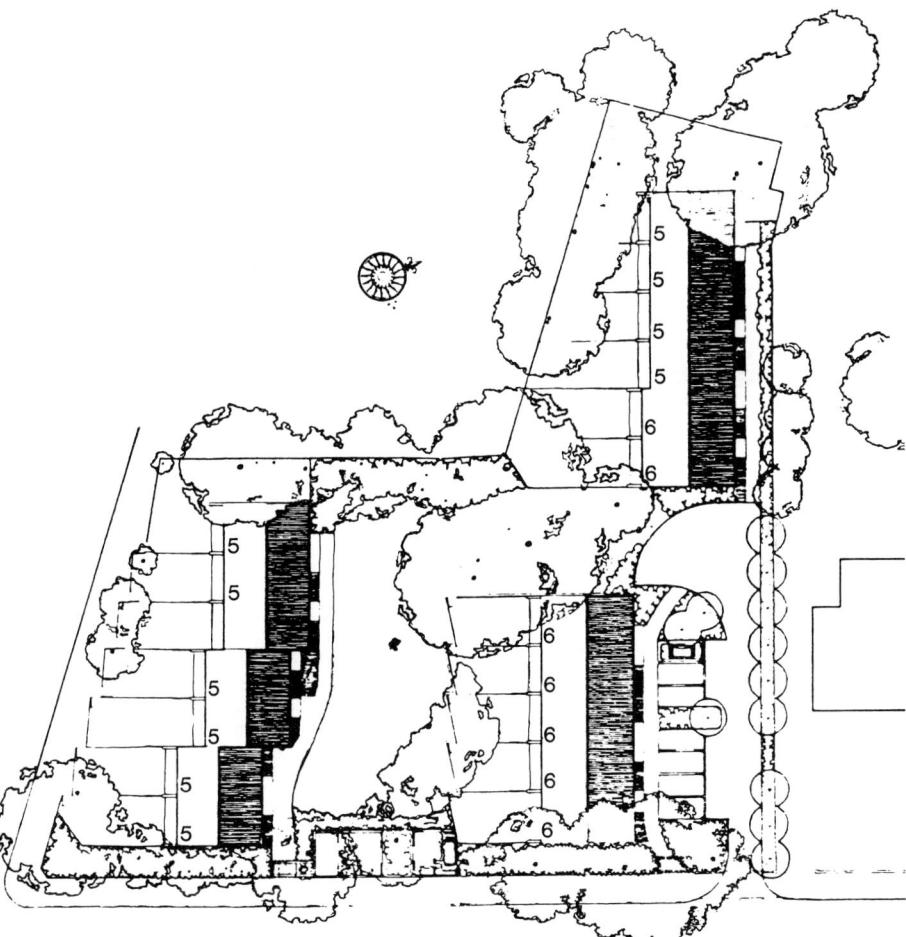

Figure 4.11. Layout of the new housing at Lawrie Park Road

determination by the architect Royston Summers and Consulting Engineers Max Fordham and Partners on how to get things done.

The environmental standards are the normal ones for English Local Authority Housing (that is, manually-operated windows – in this case, the same double sashes as at North Several – 21°C air temperature throughout for the 'Category 1' old people on ground and first floors, and 16–21°C for the young Marrieds on the second floor). A 12.5% contribution from the Active Solar Heating Installation, and 10.5% from the Passive Solar design are forecast (Fig. 4.13).

Total annual fossil-fuel consumption is estimated to be 10 kW h/m^2 of electricity, plus 123 kW h/m^2 of gas. Total delivered annual energy consumption,

Ventilation 2720W

Roof 288W

Floor 330W

Figure 4.12. Rates of heat loss

including the contribution from the Active Solar Heating, is estimated at 180 kW h/m², minimum, and 271 kW h/m², maximum. Total primary annual energy consumption has been estimated at 273 kW h/m² (including 41 kW h/m² from the Active Solar Installation).

There are thirty, two- and three-person flats in ten blocks. The entire south-facing roof area (Fig. 4.14) is occupied by water-filled solar absorbers. The absorbers have a high efficiency due to their low mass, which allows quick response, and to the low-return temperature of about 30°C, which allows efficient heat transfer. These will heat or pre-heat domestic hot water and provide partial space heating for one flat in each block from low-temperature floor coils – five

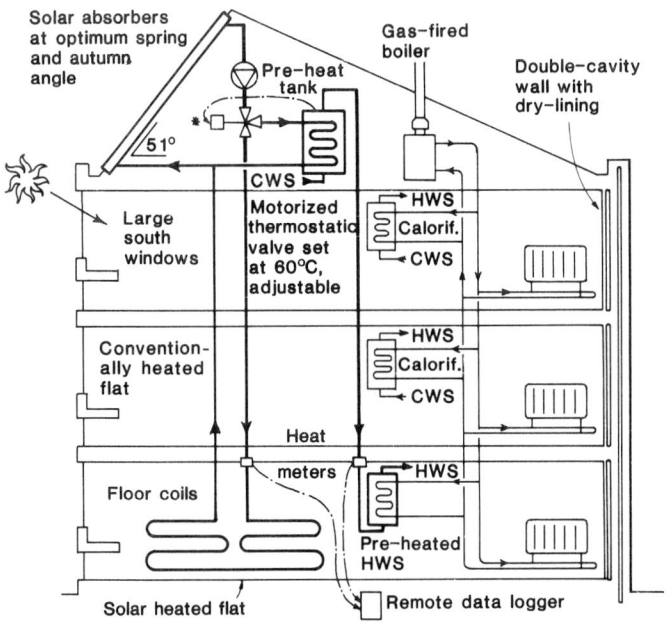

Figure 4.13. System sketch of Brownhill Road

Figure 4.14. Solar collectors at Brownhill Road, Lewisham

Figure 4.15. One flat in each block has solar-heated floor coils which help to increase the efficiency of the solar absorbers

ground-floor and five top-floor flats are heated this way to allow comparisons to be made (Fig. 4.15).

The gas-fired, back-up system is centralized, allowing the substitution of another fuel or possibly heat pumps at a later date. All parts of the system are off the shelf and installed by non-specialist subcontractors.

A users' handbook is being prepared and this will include indications of how use of the active and passive solar-heating installations will affect the heating costs. Tenants will be urged to make conscientious use of curtains to reduce heat loss through windows at night or in dull, windy and cold weather; to restrict the opening of windows and doors, and take baths when the sun shines.

Cost Yardstick limits meant that only external walls have a double cavity with normal dry lining fixed to a 150 mm Thermalite inner leaf. All fixed lights are double glazed: opening lights are single glazed and act as 'sacrificial elements' in terms of condensation, with condensation channels discharging externally; they are virtually draught proof when shut.

In order to take the greatest possible advantage of the active and passive solar-heating systems, the thermal capacity of the solar absorbers has been reduced to a minimum, while the thermal capacity of the building fabric has been kept to a maximum.

The scheme will be monitored in use for at least a year by the South London Consortium, in collaboration with the consultant engineers. The initial monitoring programme will seek to establish now much the total space heating and domestic hot water demands are met from solar sources. The monitoring will be carried out automatically, and has been designed not to interfere with the normal use of the building.

London Borough of Southwark, Wingfield Street
Wingfield Street, London Borough of Southwark (Fig. 4.16) is one of the case studies covered in this paper and is one of several small housing sites in the borough that have been collected together to form a useful number of dwellings in a programme of work. The contract has been negotiated with Llewellyn Homes Limited as part of this infill site programme. There are five-person, three-bedroom houses having a timber-framed inner skin with traditional brick cladding. The SLC Energy Group are the consultants and the objective of this project is:

(a) To minimize energy loss by passive measures without compromise of comfort levels.
(b) To supplement the residual energy requirements by low-technology solar devices.
(c) To implement (a) and (b) with maximum flexibility and control by the tenants.

The plan form accommodates the store, staircase and WC along the north 7.50 m frontage on the ground floor, whilst the living and dining rooms face south with the solar lobby acting as the link to the gardens. The perspective shows a novel measure in the use of a solar collector on the roof of the solar lobby with a simple thermosyphon system to the bathroom directly above. A prototype of this system is now under test by one of the SLC Energy Group members at Keston Kent and is providing useful results. However, there will be overshadowing by the large industrial warehouse and it is anticipated that the simple energy-saving measures, not dissimilar to those at Batley will be of value.

An interesting comparison will be made on completion between the lighter mass of this structure and the traditional double-skin masonry of other buildings.

Passive solar

The Milton Keynes area has a good record for introducing new ideas into the New Town scene. The active solar house is well documented as is the passive solar scheme at Pennylands, and the passive solar schemes at Linford.

Perhaps less well known is the Giffard Park site housing scheme commissioned by the Society for Co-operative Dwellings, with accommodation for ninety-two single people, designed by David Turrent of Energy Conscious Design, in four terraces to be built on a site adjacent to the Grand Union Canal. All terraces face

Figure 4.16. Perspective showing six new energy-saving, higher-insulation, five-person, three-bedroom houses at Wingfield Street, Southwark, now under construction with collectors above the solar lobbies employing the thermosyphon principle.

Figure 4.17. Perspective showing how the design for the proposed houses at Giffard Park, Milton Keynes, maximizes the benefits from passive solar gain

within 30° south and are planned so that all habitable rooms are on the south side. The buildings are designed to maximize the benefit from passive solar gains as can be seen on the perspective. South-facing windows are large and double glazed with low emissivity glass (U-value $= 1.6$ W/m^2 °C). Thermal capacity is provided by dense concrete blocks used in the inner leaf and party walls, whilst the first-floor construction is in pre-cast concrete.

Preliminary calculations show a 36% passive solar contribution to the annual space-heating load.

Heat pump

The introduction of heat pumps to the UK is slow compared to the United States and Germany; however, a number of trials have been carried out including schools in Essex and elsewhere; one of these is at Strawberry Hill, Wallness in Salford.

Here a low-energy housing project has been undertaken jointly by the Council and the University of Salford as one of a series of experiments being conducted to help resolve social problems by the practical application of research and technology available at higher-educational institutions.

The aim of the project is to provide housing which, whilst consuming significantly less energy, retains all the comforts and convenience of traditionally designed homes. Broadly, this is achieved in three ways: the houses are constructed with materials designed to maximize their thermal mass for heat-storage purposes; the maximum amount of insulation is provided in the cavity wall, under ground and roof space, together with dual windows; and the heat is stored and distributed either by warm-air transfer to a brick stock or by water circulation. The source of heat is an electrically driven pump which extracts heat from water tanks forming ice. The application of these design methods is illustrated in Figs 4.18 and 4.19.

Educational buildings

Schools offer excellent opportunities for the designer to contribute to the state of the art in ambient energy, and the practical work of Hampshire County Council Architects Department coupled with the research skills of the Martin Centre at the University of Cambridge has become an interesting combination.

Hampshire feel energy conservation is about the way in which people use buildings. The better their needs are satisfied and the better they understand the operation of the building, the greater is the chance that they will use the building and its equipment efficiently and in consequence, save fuel.

Coupled with this is the need, in the Local Authority field, to design buildings with low maintenance needs. Thus the sophistication of such a design depends on an elegant solution that is easily understood rather than extensive operation of automatically controlled mechanical plant.

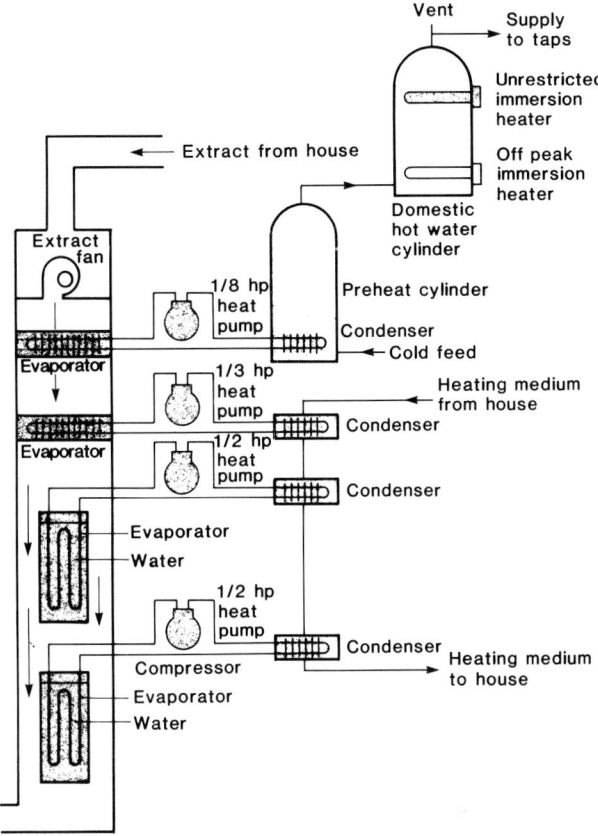

Figure 4.18. Heat-pump arrangement at the low-energy housing project at Strawberry Hill, Salford

Three two-form entry junior school designs, all to the same brief, have been studied. Two were scheduled to be built in early 1981. In one an attempt is made to use passive solar energy with some compromise on educational requirements, and in the other the educational needs have been seen to be paramount. The passive solar scheme was developed to working drawings stage before being shelved but its design has led on to the design of two other passive solar schools, one of which is scheduled to start on site in 1983.

PRACTICAL CONSIDERATIONS

This section is concerned with the practical considerations of finance, construction, maintenance, monitoring and the collection and distribution of data.

(a) For a number of years local authority housing schemes had to comply with a

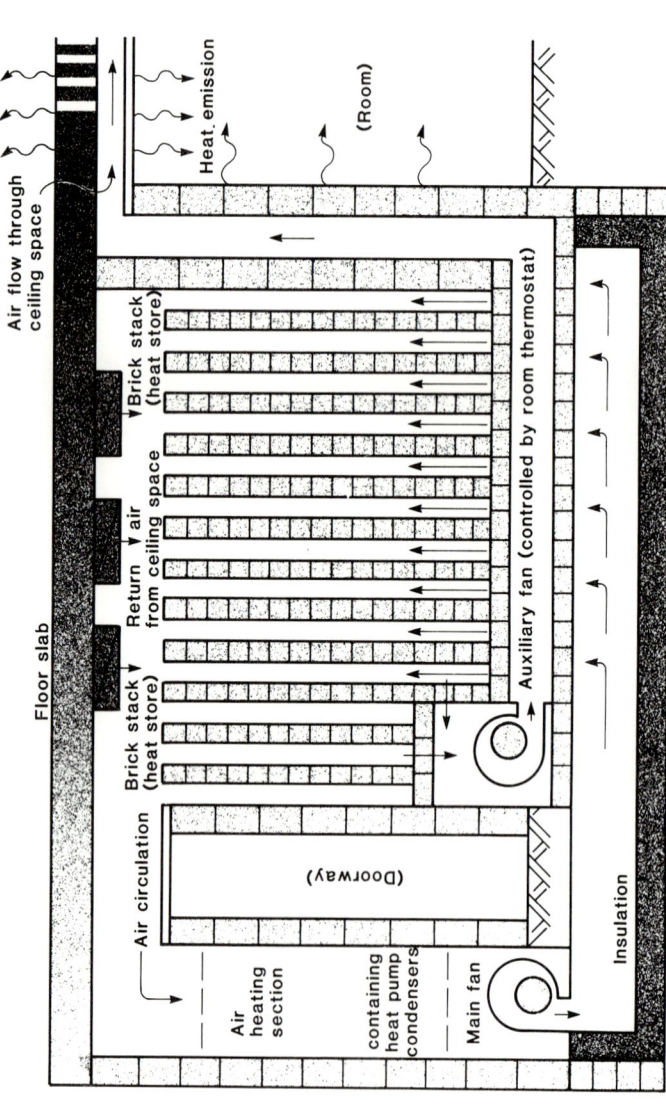

Figure 4.19. Air-circulation system in house A

series of standards laid down by the Department of the Environment, notably the Parker Morris standards. These were coupled with financial controls such as the Housing Cost Yardstick for new housing and a delegated cost limit for rehabilitation schemes.

Whilst it is true to say that any housing scheme built including ambient energy measures which met the then current Housing Cost Yardstick (HCY) criteria whether with or without the use of the 10% tolerance factor, the questions of value for money and pay back periods were not necessarily satisfied by the criteria laid down. In some cases clients demanded very high standards, whilst in other areas much lower standards were seen to be acceptable.

From 1 April 1981, the requirements will be more complex in that the Local Government Planning and Land Act under the Capital Expenditure Control will require approval for the whole project, i.e. land and buildings and demonstrations of good value for money.

The allocation of available finance to the authority will be under the Housing Investment Programme (HIP allocation) and theoretically the individual authority will be able to spend against the government allocation.

It is hoped that the greater 'freedom' under the new government proposals will result in better value for money and allow local authorities to determine their own standards, which may include ambient energy measures.

(b) The problems related to the construction are many, supervision and education being the key to success. Very high standards of supervision are vital on ambient energy schemes, anything less is doomed to failure, however well meaning the contractor. It is particularly important that full support is given by the client and quantity surveyor to the designer that condemns unsatisfactory work.

(c) Clearly, maintenance-free works are as important in the public as the private sector. Considerable effort should be made to avoid inaccessible joints, a profusion of electrical gadgetry and other electronic untried whims.

(d) Monitoring should be carried out by a professional person fully conversant with the aims of the project; too much data on an inadequate standard is already available. Good, accurate, reliable information, readily available in an understandable form is all that is needed.

(e) The distribution of data to interested parties should not be overlooked, and adquate time and money must be given to this aspect of the project.

Funding for the demonstration of energy conservation is available through a number of sources; those that are new, are related to the Commission of European Communities and the Demonstration Projects sponsored by the UK Department of Energy.

PRACTICAL PRESENTATION OF ENERGY AND ELEMENTAL COST ANALYSIS

An important point which is often overlooked in discussions about energy saving is that the initial design decisions have the greatest impact on both the energy use and the capital cost of a scheme. Once the scheme design has been finalized then the large majority of fundamental factors affecting the scheme's appearance, use, energy consumption and costs have also been determined. These factors include such items as built form, layout, orientation and often constructional materials. Once they have been determined, the detailed design stage can do very little to change these factors in anything but minor ways, and any more radical changes will result in changing the scheme design. Since the scheme design is so critical to the overall performance of the scheme it is vital to get this stage right.

For the designer to produce the best scheme design the right decisions have to be made. To do this the designer needs to know the important questions to ask at this stage, and these are normally derived from the brief. So if energy saving or whatever criterion is to be important, then it must be formulated as part of the brief either by the client, or the architect, or most probably both. However, even knowing the important questions to ask, the designer must also know how to assess how effective the sketch designs are at meeting the brief's important criteria.

The analysis techniques outlined below are intended to enable the designer to assess the effectiveness of the sketch designs as the final scheme design is worked up. These techniques are relatively traditional in nature, simple to use and, above all, applicable at scheme design stage for comparing alternative schemes, so that the most appropriate levels of energy-saving measures for a scheme's budget can be selected. This manner of working is nothing more than a commonsense principle for good building which will never change even though what can be afforded at any stage will vary with the economic climate.

As mentioned above, the analysis techniques were selected for their simplicity to use by the designer as part of the drawing-board analysis. Other people may choose to use different methods for assessing energy use and capital costs, but whatever methods are adopted, the principle of energy assessment throughout the scheme design development must always be applied if good architecture and efficient buildings are to be the end result.

The energy analysis technique is one as used, for example, by Lebens which is a slight variation on the 'simple' degree day method. The technique is based on calculating the balance point temperature for any design on a month by month basis. The balance point temperature is the outside air temperature below which the building will need to be heated to maintain the required indoor air temperature. In our calculations we have taken an average whole house internal air temperature of 18.3°C. For each design, once the average balance point temperature for the heating season has been determined, the number of degree days for the whole heating season can be established. This combined with the total thermal conductance of the building – a product of its shape, orientation, construction and insulation – will give the seasonal energy consumption requirement to maintain the required comfort conditions inside the building.

The above method has several advantages over using other degree-day methods with their standard assumptions. The extent to which the building actually needs heating is more accurately determined because the balance point temperature for each individual scheme is calculated. This enables the designer to make a more accurate allowance for improved thermal performance due to increased insulation standards and for any incidental gains from people, electrical/gas appliances and any ambient energy sources. So with this method, not only can the theoretically reduced rate of heat loss due to the energy-saving measures employed be assessed, but also the theoretical reduction of the length of the heating season. These points can be seen in Figs 4.20 and 4.21. Combining these two effects theoretically should give a much greater seasonal energy saving than is normally assumed using the 'simple' degree-day method (Figs 4.22 and 4.23).

The elemental cost analysis uses the traditional technique of pricing, using approximate quantities, each element in the sketch designs. The totals of these elemental costs will give a reasonable estimate of the scheme design's capital cost which can then be compared with the scheme's budget to assess whether that particular sketch design is economically feasible. The individual elemental costs can also be compared with the initial cost plan to identify any elements which are unduly expensive and so the designer can investigate why this is so.

Using the energy- and cost-analysis techniques together will enable the designer at the scheme design stage to select a design containing the most

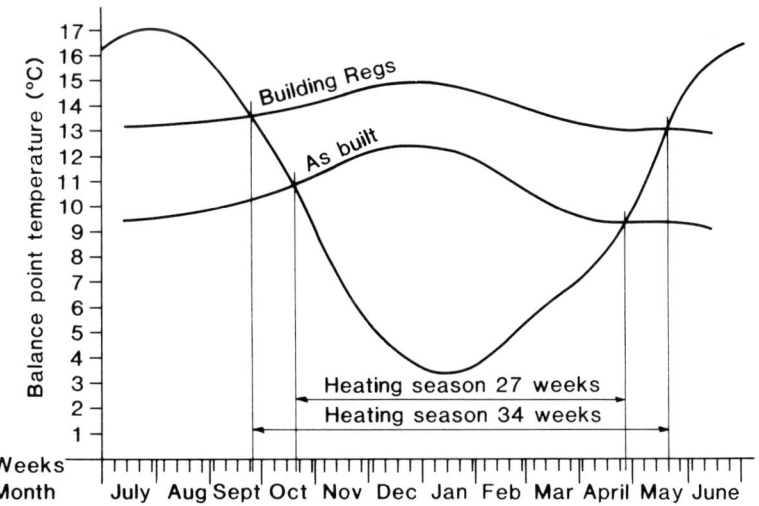

Figure 4.20. Heating season as a function of balance point temperature and seasonal temperature variation – six-person house, Askill Drive

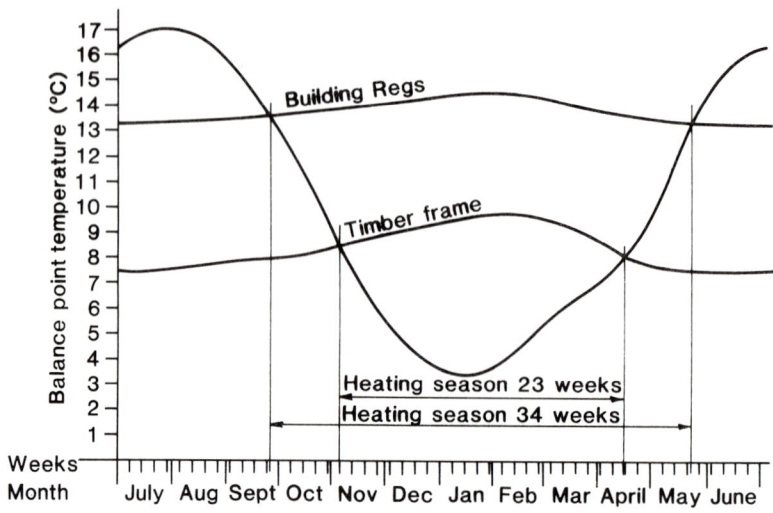

Figure 4.21. Heating season as a function of balance point temperature and seasonal temperature variation – five-person house, Wingfield Street

(a)　Seasonal heat losses

Key:　☐　House to Building Regulation Standards

　　　■　House to above Building Regulation Standards

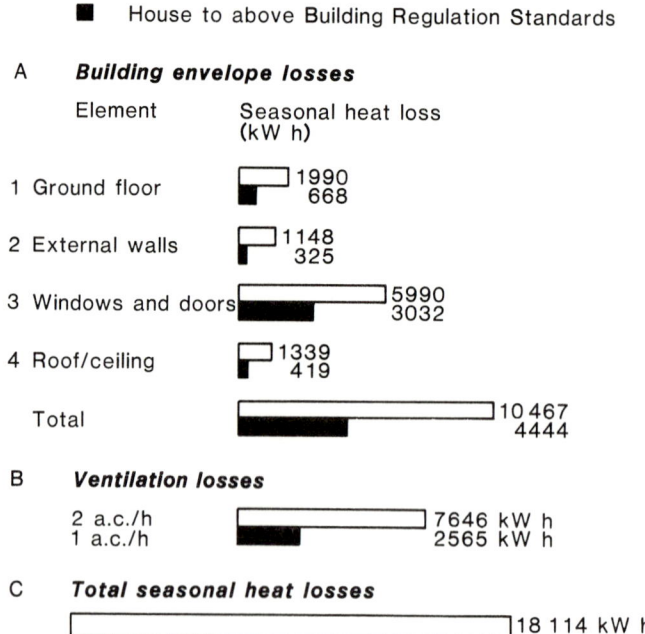

A　**Building envelope losses**

Element	Seasonal heat loss (kW h)
1 Ground floor	1990 / 668
2 External walls	1148 / 325
3 Windows and doors	5990 / 3032
4 Roof/ceiling	1339 / 419
Total	10 467 / 4444

B　**Ventilation losses**

2 a.c./h	7646 kW h
1 a.c./h	2565 kW h

C　**Total seasonal heat losses**

18 114 kW h

7 009 kW h

Figure 4.22. Six-person house, Askill Drive. (a) Seasonal heat losses; (b) capital costs (opposite)

(b)

	Element	Capital costs (£)	
1	Substructures	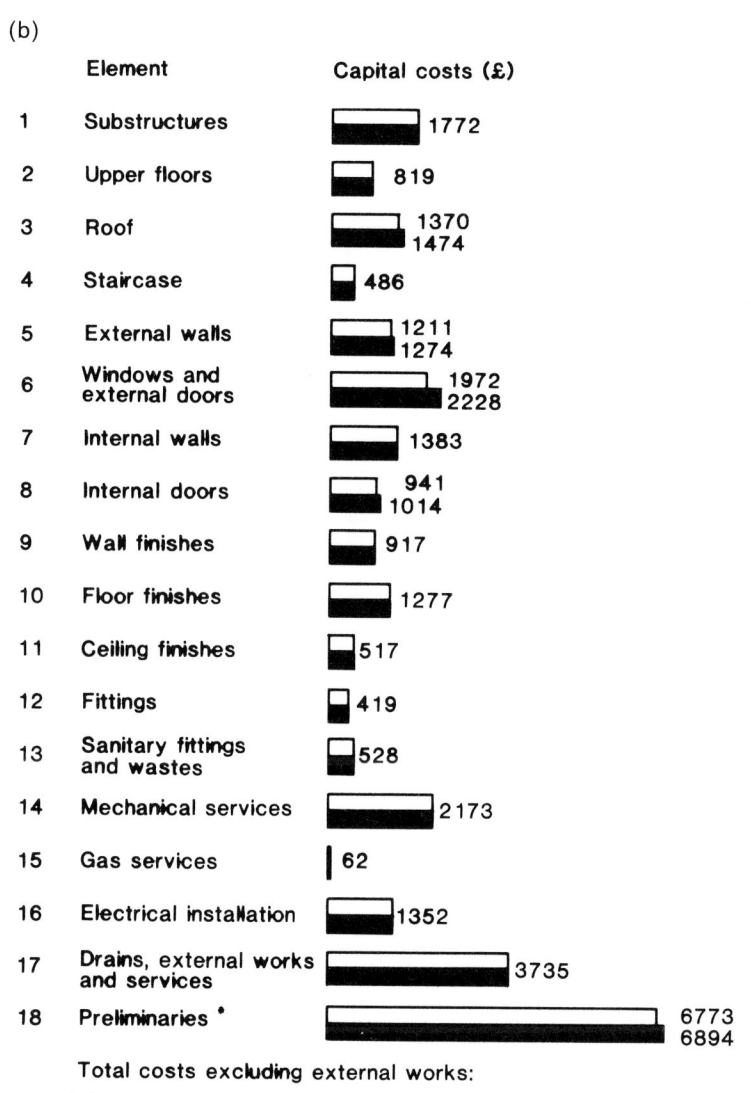	1772
2	Upper floors		819
3	Roof		1370 / 1474
4	Staircase		486
5	External walls		1211 / 1274
6	Windows and external doors		1972 / 2228
7	Internal walls		1383
8	Internal doors		941 / 1014
9	Wall finishes		917
10	Floor finishes		1277
11	Ceiling finishes		517
12	Fittings		419
13	Sanitary fittings and wastes		528
14	Mechanical services		2173
15	Gas services		62
16	Electrical installation		1352
17	Drains, external works and services		3735
18	Preliminaries *		6773 / 6894

Total costs excluding external works:

(a) House to Building Regulation Standards £27 707

(b) House to above Building Regulation Standards £28 324

(c) Budget cost for house type £26 171/£28 788

 * includes contingency sum of £828

economic energy-saving measures within the scheme's given budget. To facilitate a simple, quick and eminently visual way of analysing different designs we have presented the results of the energy and cost analysis in the form of bar charts (Figs 4.22 and 4.23). These very graphically illustrate how much the seasonal energy consumption can be reduced by simple energy-saving measures at relatively little capital cost.

To illustrate the use of these techniques we have included below a couple of case studies on two local authority housing schemes. One scheme Askill Drive, London Borough of Wandsworth which incorporated higher than the then recommended energy-saving measures, has been built within the Government Housing Cost Yardstick limits. The other scheme, Wingfield Street, London Borough of Southwark, which also incorporates higher than currently recommended energy-saving measures and also incorporates solar panels for domestic water heating, was within the HCY limits and was completed in 1981.

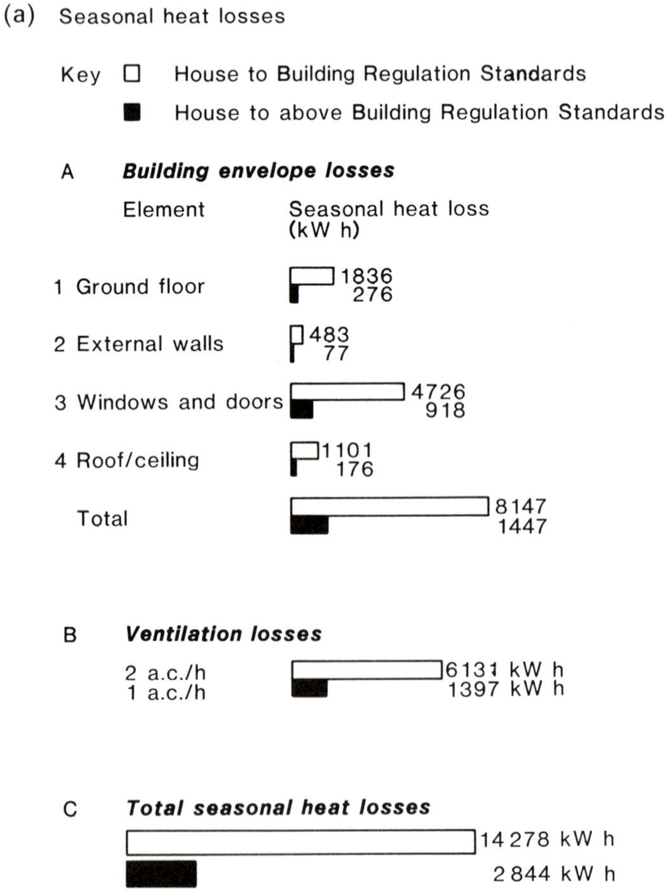

(a)   Seasonal heat losses

Key   □   House to Building Regulation Standards

■   House to above Building Regulation Standards

A   **Building envelope losses**

Element   Seasonal heat loss
(kW h)

1 Ground floor   1836
276

2 External walls   483
77

3 Windows and doors   4726
918

4 Roof/ceiling   1101
176

Total   8147
1447

B   **Ventilation losses**

2 a.c./h   6131 kW h
1 a.c./h   1397 kW h

C   **Total seasonal heat losses**

14 278 kW h

2 844 kW h

Figure 4.23. Five-person house, Wingfield Street. (a) Seasonal heat losses; (b) capital costs (opposite)

(b)

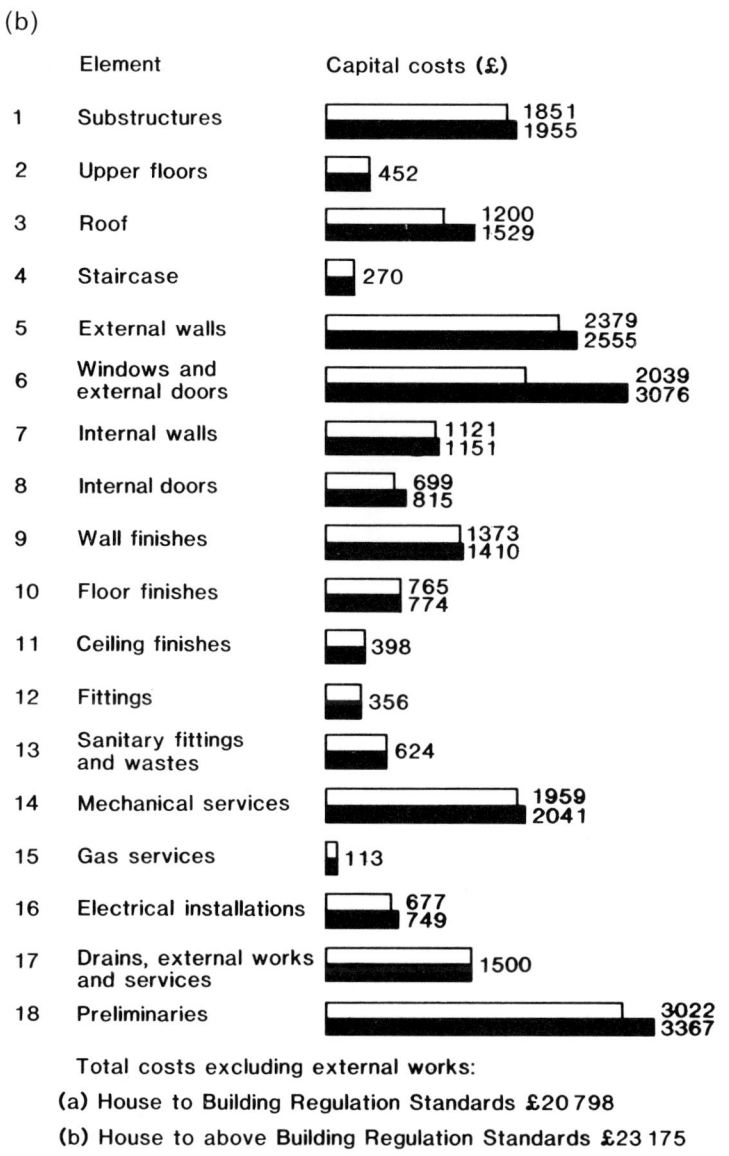

	Element	Capital costs (£)
1	Substructures	1851 / 1955
2	Upper floors	452
3	Roof	1200 / 1529
4	Staircase	270
5	External walls	2379 / 2555
6	Windows and external doors	2039 / 3076
7	Internal walls	1121 / 1151
8	Internal doors	699 / 815
9	Wall finishes	1373 / 1410
10	Floor finishes	765 / 774
11	Ceiling finishes	398
12	Fittings	356
13	Sanitary fittings and wastes	624
14	Mechanical services	1959 / 2041
15	Gas services	113
16	Electrical installations	677 / 749
17	Drains, external works and services	1500
18	Preliminaries	3022 / 3367

Total costs excluding external works:

(a) House to Building Regulation Standards £20 798

(b) House to above Building Regulation Standards £23 175

(c) Budget costs for house type £21 564/£23 720

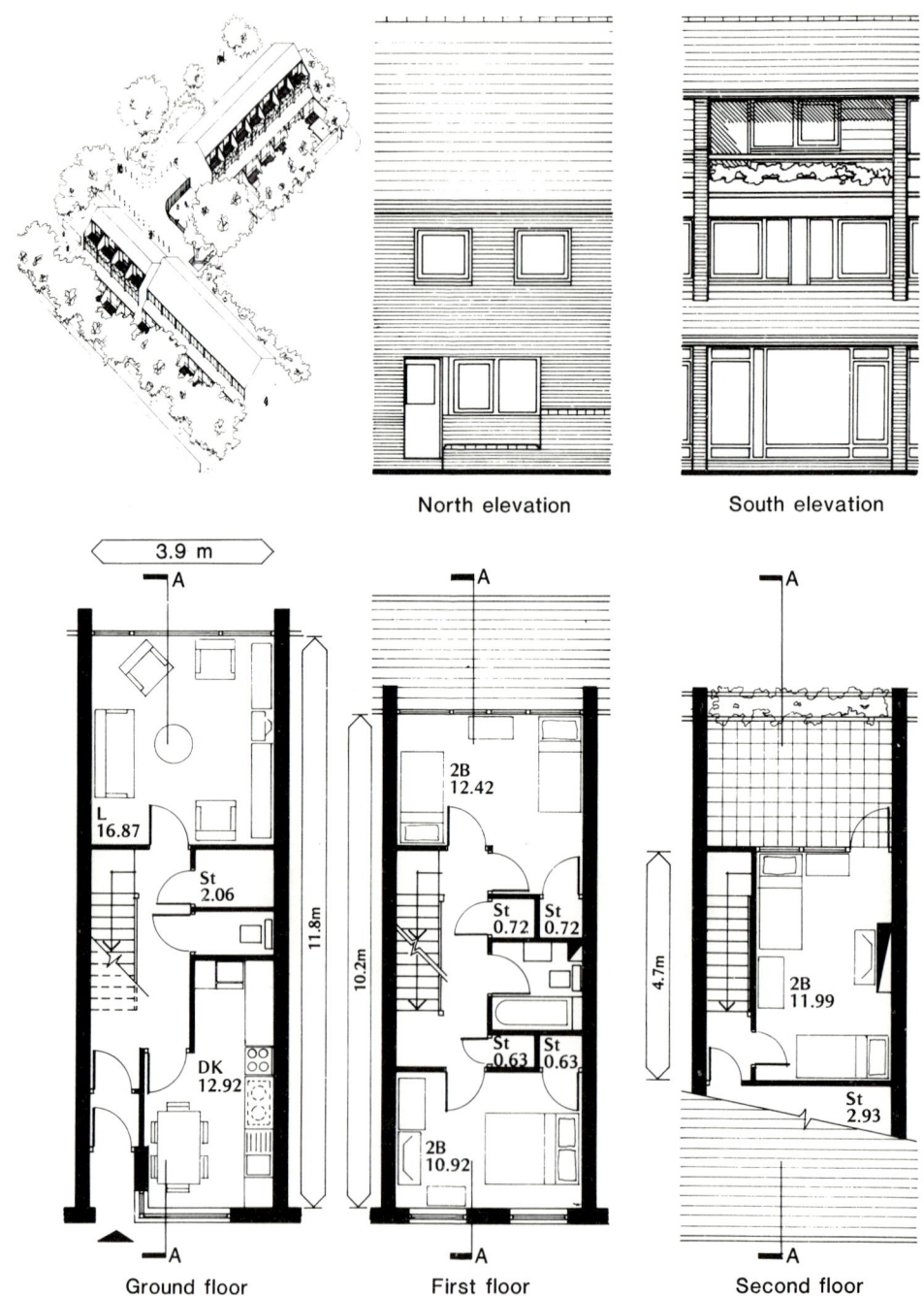

North elevation

South elevation

3.9 m

A

L
16.87

St
2.06

DK
12.92

11.8m

Ground floor

A

A

2B
12.42

St
0.72

St
0.72

St
0.63

St
0.63

2B
10.92

10.2m

First floor

A

A

2B
11.99

St
2.93

4.7m

Second floor

A

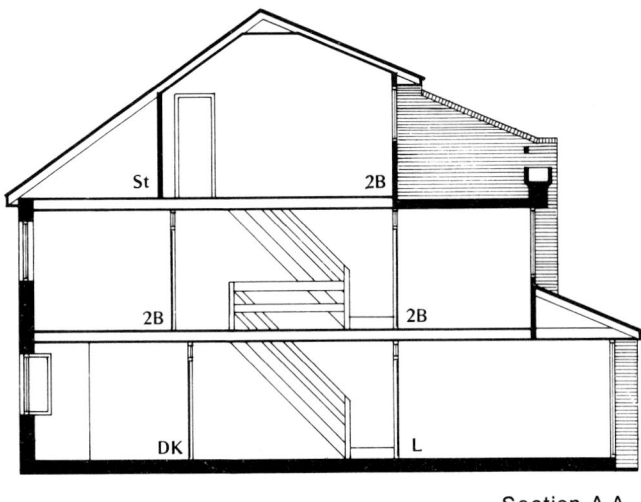

Section A.A

Figure 4.24. Askill Drive – aerial view of site, plans, elevations (see opposite) and section

Case study 1

Askill Drive, London Borough of Wandsworth, completed August, 1977 (Fig. 4.24; Tables 4.1 and 4.2). The form of the development of seven, four-person, two-bedroom, and eleven, six-person three-bedroom houses was determined by the need to achieve a relatively high density in two- and three-storey houses which would be in scale with the surroundings and provide private gardens to every house.

The intimate and secluded nature of the site was preserved by retaining all the trees in good condition. The layout of the public open space incorporates two children's play areas.

The narrow fronted (3.9 m) type plans were derived from the earlier SLC studies and included some energy-saving measures, such as draught lobbies, higher performance, preglazed, hardwood windows, mineral wool cavity-wall insulation and 100 mm roof insulation. The majority of the houses benefit from passive solar gain and one of the six-person houses has a successful domestic hot-water solar system, on which four years' monitoring data is available.

House area   104.13 m²
Parker Morris area 102.50 m²

Case study 2

The Wingfield Street site in the London Borough of Southwark (Fig. 4.25; Tables 4.3 and 4.4) referred to in the review, is typical of many small South London housing sites, being close to light industry and late-Victorian housing.

Table 4.1 Seasonal heat losses. House type: Six person – Building Regulations minimum standards, Askill Drive, LB Wandsworth

1. *Fabric seasonal heat loss*
   Average balance point temperature 14.08°C
   Degree days 1802 (to base of 14.08°C)

Element	Area (m²)	U-value (W/°C m²)	Thermal conductance (W/°C)	Seasonal heat loss (kW h)
Ground floor	46.02	1.0	46.02	1990
External cavity wall	26.55	1.0	26.55	1148
Timber cladding	–	–	–	–
Windows and exterior doors	24.3	5.7	138.51	5990
Roof/ceiling	51.58	0.6	30.95	1339
Conservatory	–	–	–	–
Total			242.03	10467

2. *Ventilation seasonal heat loss*

Volume (m³)	Rate (a.c/h)	Thermal conductance (W/°C)	Seasonal heat loss (kW h)
260	2	176.8	7646

3. *Total seasonal heat loss* (i.e. combining fabric + ventilation losses)
   (a) Thermal conductance 418.83 W/°C
   (b) Seasonal heat loss 18 114 kW h

This scheme was completed in September 1981. Weekly gas and electricity readings were started in December 1981. The site had been cleared, with a large depot on the southern aspect and a front building line on the pavement edge.

The building project is one of several that have been grouped together to form a package of small sites to enable a sound economic building programme of work to be followed.

The housing brief required six, five-person houses and the design chosen for the development is a simple terrace of wide frontage, 'single aspect' houses and was a natural response to the constraints of the site. The terrace is situated along the northern building line maximizing the site of the south-facing, private gardens to

Table 4.2 Seasonal heat losses. House type: Six person – Above Building Regulations minimum standards, Askill Drive, LB Wandsworth

1. *Fabric seasonal heat loss*
   Average balance point temperature 11.21°C
   Degree days 1209 (to base of 11.21°C)

Element	Area (m²)	U-value (W/°C m²)	Thermal conductance (W/°C)	Seasonal heat loss (kW h)	
Ground floor	46.02	0.5	23.01	668	
External cavity wall	14.5	0.45	6.53	189	325
Timber cladding	12.05	0.39	4.7	136	
Windows and exterior doors	24.3	4.3	104.49	3032	
Roof/ceiling	51.58	0.28	14.44	419	
Conservatory	–	–	–	–	
Total			153.17	4444	

2. *Ventilation seasonal heat loss*

Volume (m³)	Rate (a.c/h)	Thermal conductance (W/°C)	Seasonal heat loss (kW h)
260	1	88.4	2565

3. *Total season heat loss* (i.e. combining fabric + ventilation losses)
   (a) Thermal conductance 241.57 W/°C
   (b) Seasonal heat loss 7009 kW h

each house. This enables full advantage to be taken of the terrace's southerly orientation for solar gains. The majority of the main living areas have a southerly aspect allowing their space heating to be supplemented by passive solar gains through the south-facing windows. The conservatory roof houses the solar collectors for the domestic water-heating system which uses the thermosyphon effect; solar-heated water from the collectors circulates to the storage cylinder without needing a pump or any sophisticated electronic controls. Other energy saving measures are also included in the design.

House area   90.40 m²
Parker Morris Area 89.50 m²

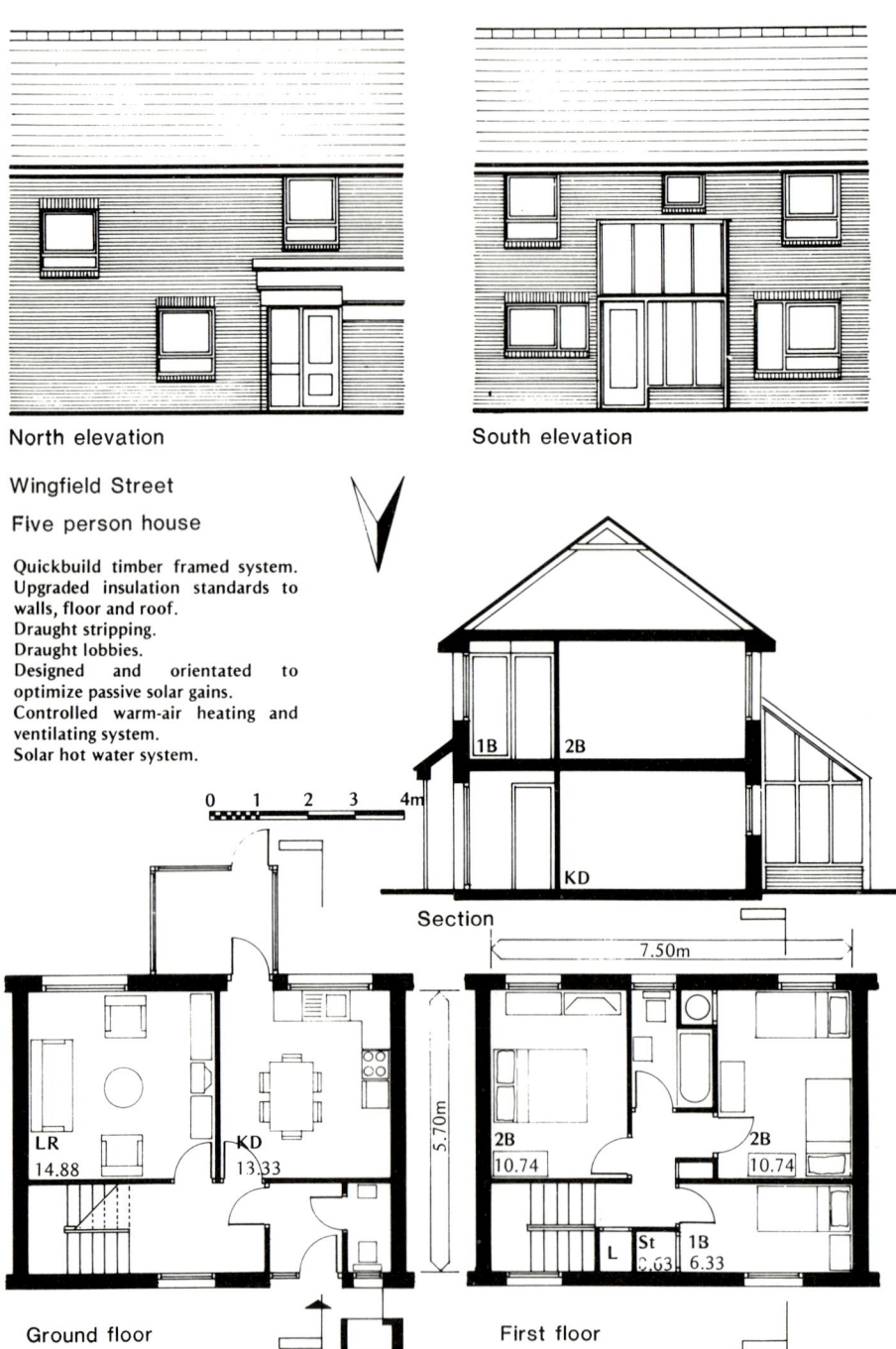

North elevation

South elevation

Wingfield Street

Five person house

Quickbuild timber framed system.
Upgraded insulation standards to
walls, floor and roof.
Draught stripping.
Draught lobbies.
Designed and orientated to
optimize passive solar gains.
Controlled warm-air heating and
ventilating system.
Solar hot water system.

0 1 2 3 4m

Section

1B   2B

KD

7.50m

5.70m

LR
14.88

KD
13.33

2B
10.74

2B
10.74

St
C.63

1B
6.33

L

Ground floor

First floor

Figure 4.25. Wingfield Street – plans, elevations and section

Table 4.3 Seasonal heat losses. House type: Five person – Building Regulations minimum standards, Wingfield Street, LB Southwark

1. *Fabric seasonal heat loss*
   Average balance point temperature 14.02°C
   Degree days 1789 (to base of 14.02°C)

Element	Area (m²)	U-value (W/°C m²)	Thermal conductance (W/°C)	Seasonal heat loss (kW h)
Ground floor	42.75	1.0	42.75	1836
External cavity wall	11.25	1.0	11.25	483
Timber cladding	–	–		
Windows and exterior doors	16.62	5.7	94.73	4067
Roof/ceiling	42.75	0.6	25.65	1101
Conservatory	9.03	1.7	15.35	659
Total			189.74	8147

2. *Ventilation seasonal heat loss*

Volume (m³)	Rate (a.c/h)	Thermal conductance (W/°C)	Seasonal heat loss (kW h)
210	2	142.8	6131

3. *Total seasonal heat loss* (i.e. combining fabric + ventilation losses)
   (a) Thermal conductance 332.54 W/°C
   (b) Seasonal heat loss 14 278 kW h

CONCLUSION

It can be seen that there have been considerable achievements in the public sector and, despite the present cuts in government expenditure, many of the schemes illustrated will continue to contribute to the fund of knowledge. How that experience is collected, recorded and distributed is of some concern and soon it is hoped that ambient energy schemes will not be seen as special projects but standard design thoroughly analysed and costed before going on site. The methods of cost and energy analysis described in this paper should enable the

Table 4.4 Seasonal heat losses. House type: Five person – Timber framed, Wingfield Street, LB Southwark

1. *Fabric seasonal heat loss*
   Average balance point temperature 8.97°C
   Degree days 815 (to base of 8.97°C)

Element	Area (m²)	U-value (W/°C m²)	Thermal conductance (W/°C)	Seasonal heat loss (kW h)
Ground floor	42.75	0.33	14.11	276
External cavity wall	11.25	0.35	3.94	77
Timber cladding	–	–	–	–
Windows and external doors	16.65	2.5	41.63	814
Roof ceiling	42.75	0.21	8.98	176
Conservatory	9.03	0.59	5.32	104
Total			73.97	1447

2. *Ventilation seasonal heat loss*

Volume (m³)	Rate (a.c/h)	Thermal conductance (W/°C)	Seasonal heat loss (kW h)
210	1	71.4	1397

3. *Total seasonal heat loss* (i.e. combining fabric + ventilation losses)
   (a) Thermal conductance 145.37 W/°C
   (b) Seasonal heat loss 2844 kW h

designer to consider energy aspects at an appropriately early stage in the design process, thus avoiding the problem of extra over-costs so frequently experienced in such projects.

FURTHER READING

Building Research Establishment (1975) *Energy Conservation; a study of energy consumption in buildings and possible means of saving energy in housing*, CP 56/75, BRE, Garston, UK, June.

Max Fordham and Partners, 57D Jameston Road, London, NW1 7DB
Telephone (01) 267 3785
Dr N. Ryding

SLC
Dr Lali Makkar

5.2 SLC Energy Group Secretary
Mary Ince

6. Passive solar

Energy Conscious Design, 11–15 Emerald Street, London WC1 3QL
Telephone (01) 405 3121 and (01) 240 0566
David Turrent, BArch, RIBA

Society for Cooperative Dwellings, 209 Clapham Road, London, SW9
Telephone (01) 737 2077

7. Heat pumps

Administrative Department, City of Salford, Civic Centre, Chovley Road,
Swinton, Manchester, M27 2AD
Telephone (061) 793 3211
Mr Mulvenna

University of Salford
Dr J. E. Randell, Senior Lecturer

Technical Services Department, City of Salford
J. M. A. Boyle, Architect

8. Educational buildings

Deputy County Architect, Hampshire County Council, The Castle,
Winchester, SO23 8UJ
Telephone (0962) 411
Derek Poole, RIBA, Deputy County Architect

The Martin Centre for Architectural and Urban Studies, University of
Cambridge, 6 Chaucer Road, Cambridge
Telephone (0223) 69501
Dean Hawkes, MA, PhD, RIBA, Director

DISCUSSION OF CHAPTER 4

T. J. Wyatt (Brown Crozier & Wyatt). In answering the question posed in his title of 'how'
to finance ambient energy sources, is Mr Kirk saying that the answer is that no finance is
needed.

Mr Kirk indicated that the average man didn't know what a simple solar system is.
Would he like to briefly give us his ideas of a simple passive system.

A. Kirk (South London Consortium). It may be a little cheeky to suggest no finance is needed but this I believe is the case. The truth is perhaps best demonstrated by a real example – a house in Askill Drive, Wandsworth – built as a passive house several years ago – and within the Government Housing Cost Yardstick – which with all its merits and demerits is the budget figure for local authority housing. This is not in itself terribly dramatic but there were difficulties that had to be overcome; for example at the time it was difficult, if not impossible to gain approval for blown mineral-wool insulation in cavities. There was also in London then a permanent ventilation bye law which required the District Surveyor to take a case to the GLC to gain a relaxation.

Another passive house built to Yardstick was in Wingfield Street, Southwark, and quite different in design concept – timber framed, light-weight. Although the possibility of additional capital expenditure would be desirable because it would allow an even better system. Askill Drive and Wingfield Street are two illustrations requiring *no* extra money.

Mr Wyatt asks for a definition of a passive system. In housing terms this simply is a design which obtains the greatest benefit from the natural environment in which it is situated. To see the many ways in which this is done requires detailed study of individually drawn designs.

J. Harrington-Lynn (Department of the Environment). The housing designs, produced by Mr Kirk appear to have achieved about 50% of their estimated energy savings by changes in their ventilation rates. Have these savings been demonstrated in practice? Particularly with the wide variation of 6 to 1 in actual energy consumption reported for housing.

A. Kirk. It is very difficult indeed to know the precise amount to which ventilation rates can be reduced. Generally, it will depend on individual orientation, wind suction and the microclimate.

A. Burke (South London Consortium). When insulation standards are increased the length of the heating season is reduced. Whether the same ventilation rate is assumed or not the ventilation loss will be reduced because heating will occur over a shorter period of time. Secondly, improving insulation means using weather-stripped windows as opposed to standard windows and in this way reducing infiltration losses. Hence the reduction from two to one air changes an hour. That assumption may be open to discussion but it is the assumption made in calculating the quoted figures.

J. Harrington-Lynn. The proposition has been made that the length of the heating season is reduced if a building is insulated. I would suggest this may not occur, particularly in commercial buildings, when pre-heating is required. The demand for pre-heating is a function of the outside temperature early in the morning, amongst other factors and therefore if the heating plant comes on at full load, even for a short time, before the building is occupied this will reduce the level of effectiveness of use of the casual gain during the rest of the occupied period. I cannot therefore see the actual heating season being reduced by the 7–10 weeks often quoted.

J. R. J. Ellis (Building Design Partnership). On the DHSS hospital study (Chapter 5) module studies have been carried out for different masses in the building. There are some areas which have intermittent heating but on the bulk of the accommodation it is continuously heated so the influence of those factors is not so significant. The DHSS adopt a policy where over the summer season the whole of the perimeter space heating is shut off except for possibly one or two critical areas.

J. Campbell (Ove Arup & Partners). In a commercial or industrial type project there will be gains into the space which will help to offset the natural heat losses. In a completely domestic environment most heat gains occur in areas like kitchens where, in attempting to vent the smells to the outside air, the gain is vented too. Unless there are some specific heat reclaim devices to allow re-use of the heat in the house the heating season will tend to stay the same length because the 'smelly' heat gains will be thrown away.

Dr H. Hörster (Phillips GmbH Forschungslaboratorium, Aachen). The heating season is reduced as a consequence of the internal gain and the external gain – the radiation through the windows.

The internal gain from appliances, people, etc. can be estimated for an average European family at 54–60 MJ/day (15–17 kW h/day). This would be stored very effectively by the normal heat capacity of the house. The radiation through the windows and the internal gain reduces the space heating demand. If the insulation is improved this demand decreases and the internal and external gains satisfy increasingly the heating requirement on days with moderate outside temperatures. As a consequence the heating season is reduced. To clarify this, in the normal house, heating requirements, if there were no internal gain, would be about 180 GJ (50 000 kW h) and the internal gain corresponds to about 20% of this. If internal gains plus gain through the windows is included the requirement comes down to 126 GJ (35 000 kW h). For the Swedish standard internal load plus solar radiation through windows provides about 50% of the heat requirement, and for the experimental house I was demonstrating, 85–90% is provided by internal gains, in consequence the heating season is reduced to just a few months (Chapter 3).

A. J. Garland (Posford Pavry & Partners). Mr Kirk mentioned a sacrificial strip of single glazing in a double-glazing window. Presumably that was to assist in condensation problems in dwellings. Did this actually work?

A. Kirk. I believe it has worked. I have seen such a small strip working well in a house at Batley, West Yorkshire, at the drying out stage.

N. S. Gamble (W. H. Gamble & Partners). How was the reduction from two to one air changes actually measured?

A. Burke. The reduction was not measured. Some nominally suggested figures of one and two air changes an hour were taken to give an idea of the reductions possible with weather stripping and a consequently reduced heating season.

N. S. Gamble. How particular buildings such as those described can have infiltration rates reduced is not really the problem but how can the construction standard of the new building be increased? How can the weather stripping be applied to new buildings generally?

A. Burke. The answer is supervision, a factor for increasing concern. We see a role for the professional not only doing the design work but also in demonstrating it, and monitoring test designs, before the decision to build a whole estate is made.

A. Kirk. Probably one of the most important points is what actually happens on site. Surprisingly little documented information is available on actual site practices. Perhaps it is the role of the architect or the engineer to train clerks of works and others. One particular problem is that when someone is very familiar with a scheme a seemingly obvious error can be missed. Another is that the sheer magnitude of checking absolutely everything makes it nearly impossible, for instance, to check that vapour barriers haven't been disturbed or damaged by the electrician.

Dr A. F. C. Sherratt (Thames Polytechnic). The Building Research Establishment have done work on ventilation rates and the integrity of the fabric as a whole in domestic dwellings. They found that of the order of a third of the ventilation of the dwelling leaving the house through the roof space was not uncommon. The air passing through holes in the ceiling holes for pipes, electric ceiling outlets, the actual loft trap – all these were allowing the ventilating air to go through into the loft and bypassing the installed insulation. Similarly, there were gaps around windows and air was leaking on pressure tests through these gaps rather more than through the crackage of the opening lights. The research covered a number of houses, most had no flues, and the average ventilation rates were of the order of 0.8 air changes/h although some were as high as 2 air changes/h and some a little lower than 0.8 air changes/h. I would suggest that air change rates of around 0.5 air changes/h are going to have to happen. It comes down to what is acceptable workmanship. A small improvement through training and supervision is all that is needed – the industry must ensure this is carried through so that tighter dwellings can be produced as standard.

J. Campbell (Ove Arup & Partners). In Sweden every house is tested. Until we do this in the UK, standards will not improve to a level which will enable infiltration rates quoted to be consistently achieved.

A. Kirk. Every drain is water tested. If the drain does not pass, the house is not allowed to be used as a habitable dwelling. The mechanics for ventilation testing are already available.

J. R. J. Ellis. Infiltration rates were considered for the hospital (Chapter 5) and also the possibility of providing controlled mechanical ventilation at minimal rates on the perimeter rooms coupled with a much tighter standard of air tightness of construction to the hospital. Some of the team members visited Sweden to investigate the high Swedish standards and how they are achieved. The most significant criteria is the level of investment in the building fabric and its construction. To achieve a tighter building is more expensive and the additional cost must be compared to the benefits provided by energy savings or is it better to invest the available money into more effective areas. Much depends on the patterns of energy distribution for a particular building.

A. Burke. Research establishments like BRE have a valuable role in identifying particular failures, and defects, whether design, construction, or both, and providing a feedback to the designer, especially when we want to be getting tighter buildings. It is important for the designer to know if it is asking the impossible of the contractor to achieve the standards of performance, such as air tightness, he specifies.

R. Marshall (Northcroft Neighbour & Nicholson). I suspect most of us would like to see the standards improve. I personally believe that it is a question of the corporate attitude of the people concerned with the project. One person alone will not achieve the desired results. The designer and the contractor and the skilled men who actually do the work must all have a pride in what they are doing. Persuasion of the team of the importance of standards is vital.

P. Jackman (BSRIA Air Infiltration Centre). Before we go headlong into making improvements to our buildings, it is necessary to have the facility to be able to predict the practical and cost effectiveness of doing this. As yet the validated tools for making these predictions are not available and part of the work of the Air Infiltration Centre is to facilitate the development of these design aids to enable reliable infiltration predictions to be made.

Dr S. J. Wozniak (Building Research Establishment). We should recognize that we are moving into an era in which standards of design and construction will be increasingly important. For example, the timber frame, low-energy house laboratories on the Garston site were fairly air-tight three months after they were built. After a year or so cracks opened up, for example between ceilings and walls. Some are large enough to account for a significant fraction of what is now a fairly high infiltration rate. This has had the result that the mechanical ventilation system installed in one of these houses has not had a chance of working properly because hot air escapes before the system can reclaim the heat.

I agree it is important to discuss methods of obtaining good workmanship on site. Some of our recent experience has shown that it is difficult to obtain good workmanship in solar water-heating systems even when the work is supervised by research scientists and consulting engineers. My opinion here is that the only way to succeed may be to have a severe financial penalty written into contracts to the effect that if work is not carried out correctly then the contractor pays for the repairs, including the time involved. We should also recognize that it is the performance of systems and buildings after many years of operation that really matters. There is an ageing effect with many technologies and it may be important not only to consider workmanship but to develop designs which, if constructed correctly, may be guaranteed to maintain their properties for a long period of time. Additionally, it may be necessary to develop tests for completed systems and building structures.

O. S. Nielsen (Property Services Agency). In Sweden as in the UK, although there are examples of very airtight houses, in many cases the air change rate at a second reading a year later has about doubled. Also we should not overlook the fact that just as important as the fabric is the way the house is being used; not only is the opening of windows significant, but also how the air is controlled inside the house. There is a limit to how finely a house can be tuned by means of the fabric alone.

J. Harrington-Lynn. Is reduction of ventilation rate in housing a good thing? An estimated 20% of housing suffers from condensation, if occupants are encouraged to close windows and stop up all the other gaps to save energy, there is a real risk the occurrence of condensation will increase.

J. Campbell. It is certainly true that nothing can be done about the ventilation without also doing something about the insulation so that the surface temperature is above dew point and even then a minimal amount of ventilation is needed.

Dr J. Twidell (University of Strathclyde). The ventilation/condensation problem is perhaps much related to social difficulties of low income. Many people cannot afford heating, especially the electrical heating so commonly available in low-income property. Thus they remain in one room, often using paraffin heating that adds to the moisture content.

Dr A. F. C. Sherratt (Thames Polytechnic). Mr Harrington-Lynn calls for higher ventilation rates, and earlier questions whether there really was a reduction in the heating season introduced by thermal insulation. I am concerned to know whether these are personal views or views of his department. One thing is very clear, we have to reduce ventilation rates if we are serious about energy conservation, and in their reduction it has to be decided first how far it can be done without running into problems, including examination of other design solutions for getting rid of the problem that might occur if low ventilation rates were imposed on a conventional house. As John Campbell said, thermal insulation has to come first and indeed it is something of a disappointment that movement towards better standards of thermal insulation has not been quicker. Even the levels of insulation currently adopted (June 1982) do not go far enough.

Mr Harrington-Lynn knows only too well the rules for dealing with condensation. It is simply to keep the temperature of the environment up. To ensure that, thermal insulation is needed or as Dr Twidell said people must be provided with fuel at a cost they can afford so that they can keep their house warm. Energy is costing more, therefore energy conservation is needed both from a cost and from other social points of view. The lower ventilation rate is a must in order to achieve energy conservation.

J. Harrington-Lynn. In my opinion the only real cure for condensation is adequate ventilation. Insulation only raises the surface temperatures and therefore the capacity of the air to retain moisture. This moisture will condense out once the heating is switched off and the surface temperatures drop, therefore the only safe approach is to provide adequate controllable ventilation and not to advocate draught stripping in a 'willy-nilly' fashion.

Departmental Policy is contained in the Building Regulations and its amendments.

Dr J. Twidell (University of Strathclyde). Surely we are discussing the wrong implications of condensation. It is not so serious that any house has fallen down because of condensation. Condensation in Glasgow houses usually results from the high cost of heat from electrical resistance heaters.

A. Kirk. Condensation does tend to be a social problem. Much housing built in the 60s was in large concrete structures with underfloor heating. Unfortunately often those who were least well endowed with money were being put into buildings which were the most expensive to heat. One of the great difficulties is that designers have gone along with the supply companies to produce systems that are suitable to them but not to the user with only a little money. The slot meter has almost disappeared for example. An option still remaining, if coal is used in housing, is that the tenant can always buy one bag of coal which to many people is preferable to a quarterly account.

Dr S. J. Wozniak. The Building Research Advisory Service is often contacted by ordinary householders who are troubled with condensation. Many extreme cases can be attributed to the use of paraffin or propane heaters where the flue gases go directly into the rooms of the building. The advice we give includes not double glazing windows in kitchens and bathrooms where moisture can be generated in large quantities. Most condensation may then occur on the window where it will do a minimum of damage.

However, providing a sacrificial element such as a piece of single glazing is not the only way of attempting to reduce the damage caused by condensation within buildings. Direct dehumidification of the air is another possible method and BRE has carried out some work on this recently.

Only a very low level of ventilation is necessary to supply sufficient air to sustain life. A more pertinent factor in some buildings may be that radon gas can be released from granite and other building materials. Obviously, there must be sufficient ventilation to prevent a significant increase in radioactivity during occupied hours. Work in the United States has confirmed that some of the foams used for cavity fill can release formaldehyde gas over a period of many years. At the present time it does not seem entirely clear whether this is a significant problem in the UK and work is continuing. However, contaminants released into the air within buildings seem set to become an important research area. Awareness is increasing as to how long on average we spend indoors exposed to a whole range of chemical contaminants.

Dr H. Hörster. In Germany our weather is similar to that in the UK and so are our problems. We believe there is one solution for new buildings which is real improved building standards coming close to those in Sweden. Controlled ventilation with heat recovery is then a must otherwise the investment in highly insulated wall construction and well-fitting windows makes no sense.

With older buildings the problem is different, but the annual air–heat losses must be taken into account. Assume an average family house in UK weather of 400 m³ with ventilation of one air change per hour. The yearly losses correspond to approximately 29–36 GJ (8000–10 000 kW h) air–heat loss which is twice the energy needed for domestic hot water for the family living there. Two air changes per hour – a realistic figure for some houses in Britain – would correspond to 54 GJ/year (15 000 kW h/year) so that approximately half the energy requirement for space heating is made up of air–heat loss. If this energy requirement is to be reduced, reduction of ventilation is absolutely necessary. The question is, to what level can it be reduced before it becomes critical.

It would be stupid not to reduce uncontrolled air ventilation in older houses for with it there is no chance to save energy. A practical solution is therefore needed and it would appear that such a solution might be controlled ventilation with heat recovery – applicable to both new and old houses or de-humidification without controlled ventilation, although in the latter case there is still the question, is the uncontrolled air exchange sufficient to satisfy the other reasons for ventilation (odour, etc.)? We see controlled ventilation with heat recovery as the preferred solution for improved older houses.

5

DHSS Low Energy Hospital Study – Extracts Relating to Energy Accounting and Solar Water Heating

PART 1   The financial implications of saving energy
                D. Allen

PART 2   The contribution of solar water heating to
                low energy hospital design
                John R. J. Ellis and M. Corcoran

INTRODUCTION

In 1979 the UK Government Department of Health and Social Security appointed a team of consultants to study the means by which energy consumption could be reduced in new hospitals. Part of the brief was to aim for an energy reduction of 50% or better and the final report, which was completed in 1983 and runs into three volumes, examines ways of achieving this objective. The two parts of this chapter are edited extracts from two sections of the report and cover aspects relating to energy accounting and solar water heating. Presented in this way they only give a brief insight into these aspects and clearly their real context can only be fully understood when read in conjunction with the remaining technical sections of the report. It is important to note that the information and comments contained in the papers do not represent DHSS design or accounting policy to be applied on their hospital projects.

A prototype 300-bed nucleus low energy hospital is currently under construction in the UK at Newport, Isle of Wight, for the DHSS and Wessex Regional Health Authority. The consultants for the Study and the prototype are Building Design Partnership – study lead consultants, building services engineers and quantity surveyors; Ahrends Burton and Koralek – architects and Gifford and Partners – structural engineers.

PART 1

The Financial Implications of Saving Energy

The brief for the study required the financial implications of the energy saving proposals to be presented in the form of an investment appraisal.

This part concentrates upon an approach, developed as part of the study, for evaluating the financial implications of energy saving proposals in general. The approach is applicable to all building types and can be adopted for all methods of energy conservation.

In describing the approach, this chapter places its use in context and highlights its advantages over other appraisal techniques.

INVESTMENT APPRAISALS

The purpose of an investment appraisal is to place in perspective the economic relationship of the costs and benefits which are anticipated to arise from a proposed investment.

The financial implications of energy saving proposals continue throughout the life of a building. This is particularly true of the annual savings of energy which arise from the proposals. It is important, therefore, that the energy accounting approach adopted can accommodate and compare all costs and benefits which will occur over the life of a building. This requirement precludes the 'simple pay-back' approach, which only takes account of those costs and benefits occurring prior to pay-back being achieved.

An existing and well-established approach, known as 'life cycle costing', enables the necessary comparisons to be made. The approach requires the costs and benefits to be estimated both as to amount and timing. They are then converted into common time and common money terms by use of a supporting technique known as 'discounting'.

Discounting recognizes that money currently in hand can be invested to earn a return. Discounting provides the means by which one can answer the question: 'How much would I have to invest now at a given rate of interest to generate a required sum at some future date?'

After discounting, the investment costs and benefits are expressed in present value terms. Depending upon how they are related, they provide alternative measures of economic efficiency. These alternative measures are usually referred to as 'discounted pay-back period', 'net present-value investment benefits', 'internal rate of return', 'benefits-to-costs ratio' or 'savings-to-investment ratio'.

In the public sector, life-cycle costing is the recognized approach to investment appraisals. The measures adopted are usually the 'internal rate of return' or 'benefits-to-costs ratio'. For an investment proposal to be acceptable, it usually has to satisfy specified investment criteria and be affordable within the capital monies available. Where capital is limited, alternative proposals normally have to be 'ranked' to establish their relative priority for capital allocation.

Costs and benefits, respectively, can be the net result of a series of individual positive or negative costs and positive or negative benefits. The question arises: 'Is a negative cost a benefit?' or conversely: 'Is a negative benefit a cost?' The answer depends upon the context and the objective. For instance, in a context where the objective is reducing running costs, benefits are measured in terms of the running costs saved. In the same context, costs are measured in terms of the other financial implications of saving running costs (i.e. capital costs).

Even if there are no limitations as to the amount of capital monies available for investment, it is still necessary to distinguish between costs and benefits to enable comparisons to be made with investment criteria.

APPRAISING ENERGY SAVING INVESTMENTS

In an energy saving context, where the objective is saving energy, benefits are measured in terms of the energy costs saved. Costs are normally measured in terms of all other financial implications of saving those energy costs. In other words; 'How much does it cost to save energy?'

The measure of benefits in this context is different from the measure appropriate to a running cost saving context. In the energy saving context, the measure is the saving of energy costs. In the running cost saving context, the measure is the saving of running costs. These clearly are two different measures and can result in different ranking of the same proposals.

It is normal for benefits to be evaluated in financial terms. According to the Department of Energy in their publication *Energy Trends* [1], the real price of (industrial) fuels has been increasing at a greater rate than other costs since 1973. It seems likely that a similar pattern will continue, but for how long and at what rate is unknown. This presents design teams with difficulties in estimating the financial benefits of energy saving investments. The financial benefits assumed will often determine the acceptability of a proposed investment.

The combination of the design team's difficulty in estimating the financial benefits of saving energy and the absence of energy-sensitive investment criteria tends to militate against the saving of energy. This tendency is often further exacerbated by the confusion which exists as to the true objective of the appraisal. Is it to save energy or reduce running costs?

Given the objective of saving energy, what is needed is a new approach, specially devised for the appraisal of energy-saving investments. The approach must recognize the need for the client's investment criteria to be sensitive to and dependent upon the benefit attributed to the saving of energy.

A NEW APPROACH

The approach adopts life-cycle costing, but separates the traditional components into two groups:

(a) Those components which are within the design team's competence to assess.
(b) Those components which the client is required to determine.

Design teams can assess the annual quantity of energy saved by a proposal. They can also estimate the non-fuel (i.e. capital and non-fuel revenue) cost implications of a proposal which will occur over time. After discounting, those estimates can be combined with the assessments of energy saved. The resultant expression is in terms of the number of units of energy saved per net present value pounds sterling (£NPV) spent as a consequence of the energy saving investment proposed. This has been called the *energy conservation investment factor* (ECIF), which is expressed in terms of kW h/annum/£NPV spent.

The client's considerations need to result in investment criteria expressed in the same terms as the ECIF. This has been termed the *energy conservation investment yardstick* (ECIY).

ECIF can usefully be thought of in terms of the energy saved and ECIY in terms of the energy to be saved. Both are expressed in the same units. A simple comparison of ECIF and ECIY determines the acceptability of a proposal.

Example
An energy saving proposal in which the estimated annual saving of energy is 100 000 kW h and the net present value spent is £NPV 7000 would have an ECIF of:

$$\frac{100\ 000\ kW\ h}{£NPV\ 7000} = 14.3\ kW\ h/annum/£NPV\ spent$$

Given (say) an ECIY of 9 (kW h/annum/£NPV spent), this particular proposal would be acceptable. It saves 14.3 (compared with 9) kW h/annum/£NPV spent.

Had the ECIF been less than 9 (e.g. if the proposal had an NPV of £12 000 – and, therefore, an ECIF of 8.3) it would not be acceptable. It would only save 8.3 (compared with 9) kW h/annum/£NPV spent.

THE CLIENT'S CONSIDERATIONS

In essence, the whole of the client's considerations relate to the financial benefits of saving energy. The question – 'What present value do I place on saving one unit of energy per annum over the life of the building?' – needs to be answered.

To answer this question, consideration needs to be given to the following: the building's life (including its likely operational start date); the discount rate to be adopted; and the likely price of energy over time relative to other costs. Each of these has a significant effect on the ECIY which results.

The building's life has a direct effect on the number of units of energy taken into account. Disregarding year-to-year variations, this can be taken as a linear relationship as shown in Fig. 5.1.

Discount rates can have a dramatic effect on the viability of investments. Lower discount rates (see Fig. 5.2) result in greater present values. This means that the value of energy savings made during the building's life is increased.

Longer building life, lower discount rates and higher energy prices each contribute to lower ECIYs. Lower ECIYs result in more energy saving proposals being acceptable and more energy being saved. These effects are all statements of arithmetic fact. The actual building life, discount rates and future energy prices adopted are matters of judgement, assumption and, in the case of energy prices, speculation.

It is generally accepted that a sixty-year operational building life should be adopted for investment appraisal purposes. In fact, buildings often last much longer. The admission of longer building life into the appraisal would further benefit the saving of energy.

Discount rates vary. In the UK public sector, real rates of 5% and 7% are adopted. In the USA public sector, a constant (real) rate of 10% is used. Real discount rates are only appropriate where the effects of general inflation are excluded from estimates of costs and benefits. According to the HM Treasury booklet *Investment Appraisal and Discounting Techniques and the Use of the Test*

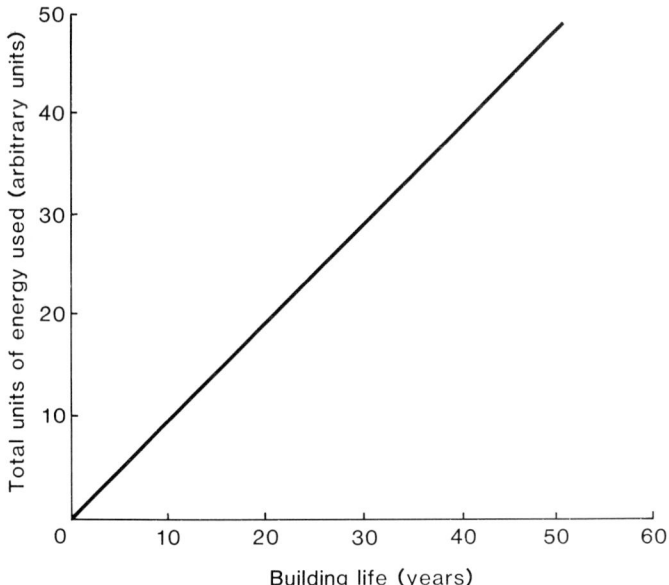

Figure 5.1. Building life/units of energy

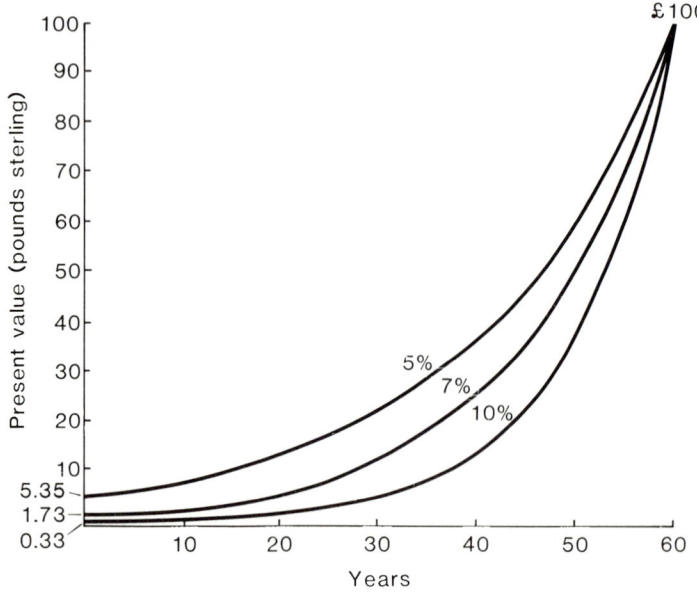

Figure 5.2. Effect of different discount rates

Discount Rate in the Public Sector [2] paragraph 9: '... 5% may be used; (a) where it can be shown by experience that the assessment of costs and benefits can be made with good degree of objectivity, and (b) when the problem is one of simply comparing different techniques of production for meeting a given output'.

For both these reasons, it can reasonably be argued that the lower rate should always be adopted when appraising the economics of energy saving proposals. Depending upon the view taken of energy prices, there could also be good reason for reducing the discount rate below 5%. In the private sector, it is for the client to determine the discount rate to be adopted, with guidance from the design team as appropriate.

The matter of energy prices warrants special and separate consideration.

CONSIDERING THE REAL PRICE OF ENERGY

It has already been said that the building's life has a direct effect on the number of units of energy taken into account. Each of those units has to be allotted a price. In reality, nobody knows now what the real price of energy will be over a building's life. The Department of Energy, in its publication *Energy Projections 1979* [3], adopts two possible scenarios, but states:

'The projections ... examine possible UK energy demand and supply prospects to the end of the century. They are necessarily based on certain broad long term assumptions about economic growth (which do not exhaust the range of possibilities), technical improvements and movements in energy prices, and are therefore not predictions of what will necessarily happen nor prescriptions of what should happen. The results would, of course, vary if different assumptions were used.'

Those projections only extend to the year 2000. Adopting a sixty-year building life, current energy saving considerations need to extend beyond the year 2040.

The Department of Energy's projections and the reality of the last seven or eight years tend to suggest a continuing upward trend in the real price of energy. The question remains: 'For how long and at what rate?' There is little specific guidance as to the answer to that question. It might also be appropriate to consider the present value of energy from a different point of view.

Prices of fuels will progressively reflect their limited availability. Government may intervene and accelerate this process. By its intervention, prices will increase and the rate of consumption will tend to slow down. A more immediate effect could be achieved by giving each fuel type a conservation value for investment appraisal purposes. The approach proposed could readily accommodate such conservation values. The effect would be to encourage conservation of those fuels which are least plentiful. In this way, consumption of limited fuel supplies could be slowed down well ahead of the influence of prices. A more realistic view might be to consider the affordability of conserving energy.

The viewpoint adopted and the value placed on saving energy will vary between clients. That viewpoint is likely to have significant implications on the amount of energy consumed and conserved over the years.

In Figs 5.3–5.6, four different scenarios are depicted. Whilst no particular significance should be attached to the scenarios shown, they do serve to illustrate the effect of different fuel price assumptions.

(a) Scenario A, shown in Fig. 5.3, depicts real fuel prices doubling by the year 2000 and continuing at a similar rate thereafter.

(b) In scenario B, shown in Fig. 5.4, coal, gas and electricity have been assumed to double and oil to quadruple in price by the year 2000 and continue at a similar rate thereafter.

(c) Scenario C, shown in Fig. 5.5, adopts the primary conversion factors given in Building Research Establishment Current paper CP 56/75 [4]. These are reflected in the relationships depicted between the value of each fuel type. The remainder of the scenario is based on the real price of oil doubling by the year 2000 and continuing at a similar rate thereafter.

(d) Scenario D, shown in Fig. 5.6, adopts assumed conservation values. The relationships between the values of each fuel type reflect their assumed relative scarcity. The remainder of the scenario is based on the real price of oil doubling by the year 2000 and continuing at a similar rate thereafter.

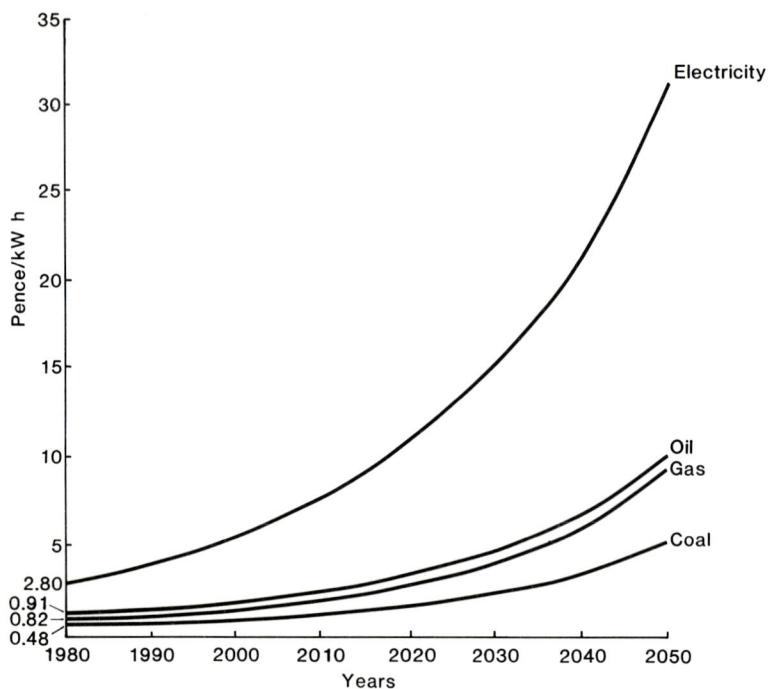

Figure 5.3. Fuel price scenario A – all fuels double in price by the year 2000

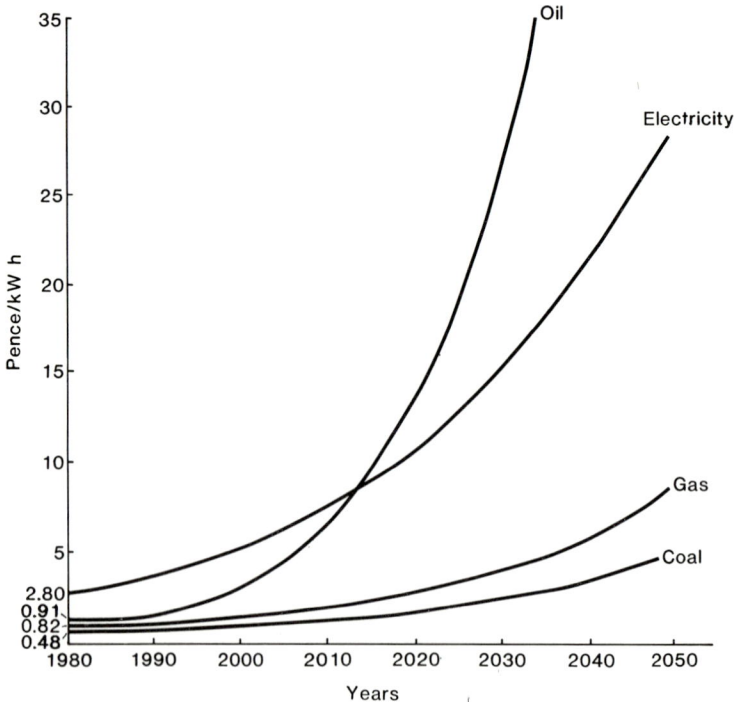

Figure 5.4. Fuel price scenario B – oil prices increase at twice the rate of those of other fuels which, as in scenario A, double by the year 2000

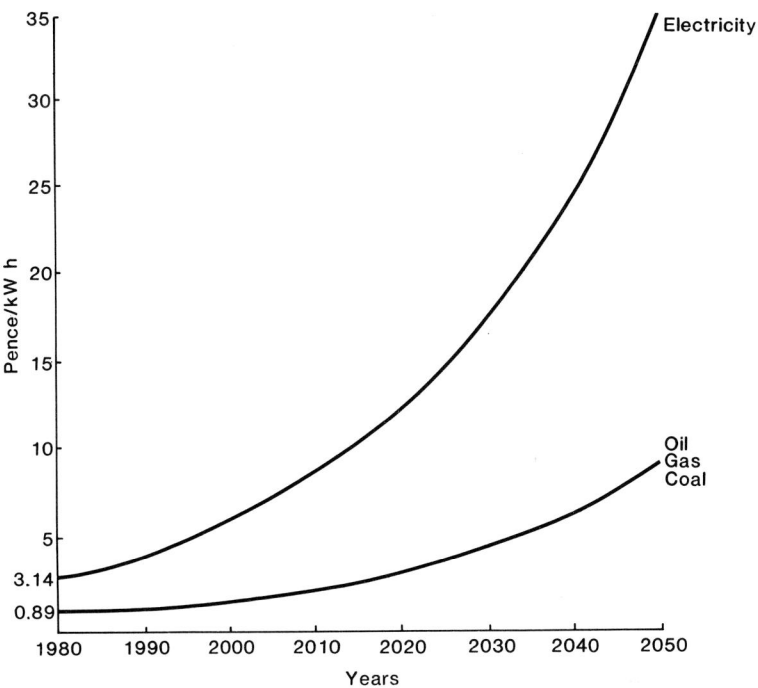

Figure 5.5. Fuel price scenario C – the primary fuel conversion factors suggested by the Building Research Establishment are assumed

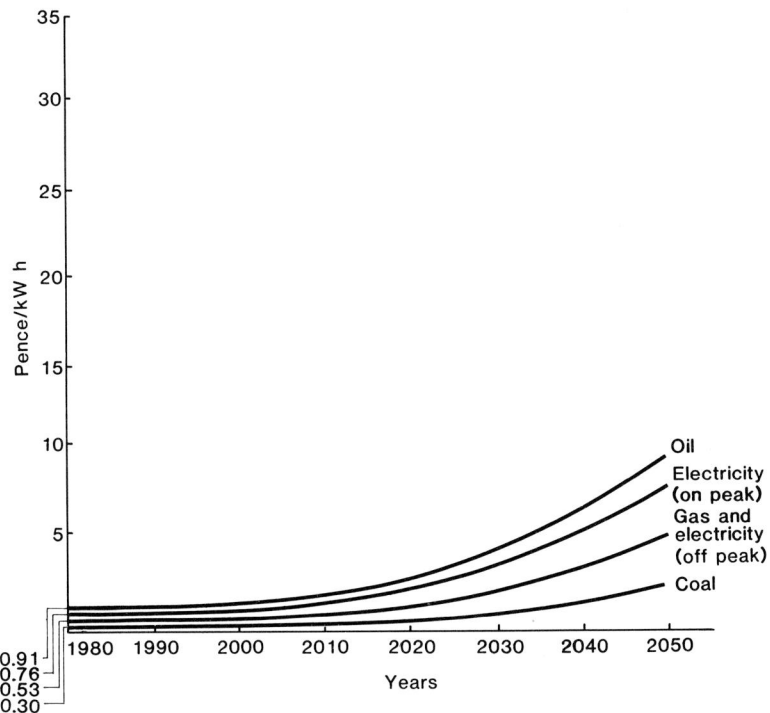

Figure 5.6. Fuel price scenario D – assumed conservation values are adopted by adjusting the 1980 price to reflect an assumed relative scarcity

MIXED FUEL CONSIDERATIONS

For the purpose of simplicity, the method of dealing with mixed fuel energy saving proposals has not so far been contemplated. Movement in the consumption of more than one fuel type as a consequence of an energy saving proposal is accommodated by converting each fuel into a single fuel equivalent. The choice of that single fuel is entirely optional.

Adopting coal, the ECIY becomes an expression of the present value of saving coal over the life of the building and conversion of fuels into 'coal equivalents' is achieved by factorizing. The factors used would be the arithmetic relationships of the present value allotted to each of the other fuels compared to that of coal. It will be noted that the coal equivalents referred to here are a different measure to those known as 'million tonnes coal equivalent' (Mt c.e.).

Example

An energy saving proposal in which the estimated annual energy consumption implications are: a saving of 300 000 kW h of electrical energy but with an increase of 200 000 kW h of gas energy and the net present value spent is £NPV 70 000.

Given conversion factors of, say:

> Coal: 1.00
> Gas: 2.00
> Oil: 2.00
> Electricity: 6.00

The factorized energy implications would be:

> Gas: $+200\,000$ kW h $\times 2.00 = +\ \ 400\,000$
> Electricity: $-300\,000$ kW h $\times 6.00 = -1\,800\,000$

Annual saving of coal equivalent units of energy is 1 400 000 kW h.
The ECIF of the proposal, therefore, would be:

$$\frac{1\ 400\ 000 \text{ kW h}}{\text{£NPV } 70\ 000} = 20.00 \text{ kW h/annum/£NPV spent}$$

Given, say, an ECIY of 9 (kW h/annum/£NPV spent), this particular proposal would be acceptable. It saves 20.00 (compared with 9) kW h/annum/£NPV spent.

In order to determine the conversion factors and the ECIY, it is necessary first to calculate the present value of saving 1 kW h of energy per annum over the life of the building. A separate calculation has to be carried out for each fuel type under consideration.

Having considered and established the likely price of energy over time, the price of 1 kW h of energy in each year of the building's life can be readily determined. Applying the selected discount rate, each year's price can be discounted back to base data and expressed in present value terms. The sum of those separate values represents in a single figure the present value of saving one unit of energy per annum over the life of the building. The present value of each fuel type under consideration can be calculated in this way.

The ECIY is the reciprocal of the present value of the fuel type selected as the single fuel equivalent. In other words, the present value for a particular fuel is the value of saving 1 kW h of it per annum over the life of the building. Conversely, the ECIY expression is the number of units of the selected fuel required to be saved per annum over the life of the building for each £NPV spent on capital and non-fuel running costs.

Table 5.1 shows the present values, ECIYs and conversion factors of each of four fuel types in a range of time horizon, discount rate and energy price scenarios. The scenarios are those depicted on Figs 5.3–5.6 and the single fuel adopted in each case is coal. Each time horizon includes a six-year design, construct and commission period. Values have been presented as calculated. Rounding of these values is not recommended as this could produce significant distortions when applied to large quantities of energy.

THE DESIGN TEAM'S REQUIREMENTS

In an energy saving context, an essential component of each design team's brief is its energy brief. That brief needs to include a clear statement of the client's energy saving investment criteria. They may be prescribed by the client or suggested by the design team and agreed to or modified by the client.

Adopting the new approach and to ensure compatibility of ECIFs and ECIYs, the energy brief needs to include:

(a) *Base date:* the common time datum to which all costs and benefits are converted
(b) *Inflation assumptions:* clarifying whether or not the effects of general inflation are to be included
(c) *Discount rate:* the annual percentage rate adopted in discounting the investment costs and benefits back to the base date
(d) *Building life:* that component of the time horizon which relates to the operational life of the building
(e) *Conversion factors:* the factors to be applied to each fuel to convert it into single fuel equivalents
(f) *The ECIY:* to provide a yardstick against which the ECIF of each energy saving proposal or selection of proposals may be compared

Table 5.1 Present values, ECIYs and conversion factors for the various fuels

Fuel type:	Coal				Gas				Oil				Electric			
Time horizon:	26 years		66 years		26 years		66 years		26 years		66 years		26 years		66 years	
Discount rate:	5%	7%	5%	7%	5%	7%	5%	7%	5%	7%	5%	7%	5%	7%	5%	7%
Scenario A																
Present value pence	7.70	5.83	17.82	10.38	13.16	9.97	30.44	17.74	14.61	11.07	33.78	19.68	44.95	34.06	103.96	60.58
ECIY	12.99	17.15	5.61	9.63	12.99	17.15	5.61	9.63	12.99	17.15	5.61	9.63	12.99	17.15	5.61	9.63
Factor	1	1	1	1	1.71	1.71	1.71	1.71	1.90	1.90	1.90	1.90	5.83	5.83	5.83	5.83
Scenario B																
Present value pence	7.70	5.83	17.82	10.38	13.16	9.97	30.44	17.74	24.52	18.22	112.47	54.78	44.95	34.06	103.96	60.58
ECIY	12.99	17.15	5.61	9.63	12.99	17.15	5.61	9.63	12.99	17.15	5.61	9.63	12.99	17.15	5.61	9.63
Factor	1	1	1	1	1.71	1.71	1.71	1.71	3.18	3.13	6.31	5.28	5.83	5.83	5.83	5.83
Scenario C																
Present value pence	13.80	10.46	31.93	18.60	14.29	10.82	33.04	19.25	14.61	11.07	33.78	19.68	50.41	38.19	116.58	67.93
ECIY	7.25	9.56	3.13	5.38	7.25	9.56	3.13	5.38	7.25	9.56	3.13	5.38	7.25	9.56	3.13	5.38
Factor	1	1	1	1	1.03	1.03	1.03	1.03	1.06	1.06	1.06	1.06	3.65	3.65	3.65	3.65
Scenario D																
Present value pence	4.81	3.64	11.13	6.49	8.51	6.44	19.67	11.46	14.61	11.07	33.78	19.68	12.20	9.24	28.21	16.44
ECIY	20.79	27.47	8.98	15.41	20.79	27.47	8.98	15.41	20.79	27.47	8.98	15.41	20.79	27.47	8.98	15.41
Factor	1	1	1	1	1.76	1.76	1.76	1.76	3.04	3.04	3.04	3.04	2.54	2.54	2.54	2.54

RANKING PROPOSALS AND SELECTION

The ECIF can be used to rank a selection of energy saving proposals where capital monies available for investment are limited. The resultant ranking is in order of the most energy saved per annum per £NPV spent (i.e. those which provide the greatest benefits).

Capital costs are only part of the £NPV spent component of the ECIF. Therefore, to enable the benefits to be optimized within the capital monies available, it is necessary to identify the underlying capital cost component of each proposal. Optimization is achieved by adopting that selection of proposals which, within the limitation of the capital monies available, results in the greatest number of kilowatt hours of energy being saved per annum per £NPV spent.

Example
An energy saving context is present in which the following hold:

(a) The proposals shown in Table 2 are under consideration.
(b) The client's ECIY (for each proposal) is 4.5 kW h/annum/£NPV spent.
(c) The maximum capital monies available for energy saving measures are £100 000.

Table 5.2 Ranking

Proposal	Energy saved (kW h/annum)	ECIF	Capital spent (£)	
			Individual	Cumulative
1	200 000	11.5	15 000	15 000
2	330 000	6.5	80 000	95 000
3	350 000	5.0	85 000	180 000
4	105 000	0.9	150 000	330 000

The proposals have been ranked in order of the most energy saved per annum per £NPV spent (i.e. by their ECIFs). Their capital cost implications have been stated and accumulated in order of their ranking.

Given the client's ECIY of 4.5 (kW h/annum/£NPV spent), each of the first three proposals would be acceptable. However, given also the capital constraint of £100 000, only a combination of the first two or of the first and third proposals would be acceptable.

Either of these combinations is acceptable in relation to the ECIY and can be afforded within the £100 000 capital constraint. The third proposal saves more energy per annum than the second, but its ECIF is less favourable. Its non-fuel revenue implications must, therefore, also be less favourable.

Assuming no further client constraints or preferences, either proposals 1 and 2 or 1 and 3 could be adopted. In practice, the client's view of such alternatives would probably be sought.

The client's ECIY may be set for comparison with each proposal or, alternatively, a collection of proposals. Where it is set for the latter, there is a tendency for more proposals to be acceptable. Those which, individually, have ECIFs in excess of the ECIY, when aggregated, generate a collective ECIF in excess of the ECIY. This facilitates the inclusion of additional proposals which, individually, have ECIFs lower than the ECIY. However, the degree to which this applies depends upon the level of capital constraint.

ADVANTAGES OF THE NEW APPROACH

The main advantage of the new approach is that it is specifically designed for energy saving considerations. It leaves no doubt as to what constitutes a benefit and highlights the need for an energy brief.

In highlighting that need, it requires the client to express his investment criteria in terms of his view of the present value of saving each type of energy. In so doing, it serves to underline the need for investment criteria and energy prices to be considered interdependently.

As an approach, it retains the benefits of the IRR (internal rate of return) and BCR (benefits-to-costs ratio) approaches. In an energy saving context, it is a direct alternative to the BCR approach. All other things being equal, the same set of proposals will rank identically adopting either the BCR or the ECIY approach.

The approach facilitates the evaluation of energy saving benefits in terms of value as well as price. This provides scope for directing investment decisions towards the conservation of particular forms of energy by giving each fuel type a weighted value.

The public sector could adopt a single ECIY and a single set of conversion factors, both set by the Treasury for use on all public sector projects. It might also be possible for the Government to influence the ECIYs and conversion factors set by private sector clients. This could be achieved via guidance material or direct incentive introduced via the ECIY approach.

Investment appraisals are primarily a method of accounting and comparing alternative proposals. Adopting the ECIY approach, the conversion factors can also provide a useful guide to designers. When considering alternative methods of saving energy, design teams need to have in mind the relative values of the alternative energy sources at their disposal. Traditionally, current price relationships tend to be taken as the guide. Adopting the conversion factors provided as part of the energy brief, present value relationships would provide the necessary guidance.

The present value of each fuel type could also be adopted in selecting the fuel mix for a project. Fuel choice is normally based on current fuel prices. The energy brief could be extended to include a statement of the present value to be adopted for each fuel type. This would enable fuel policies to be based on the same considerations as those which apply to energy saving measures.

SUMMARY AND CONCLUSIONS

Investment decisions are a key factor in the saving of energy. They are determined by the view taken of the present value of saving energy. In the absence of known future energy prices, investment decisions are often based on current prices and relatively short pay-back periods. The long term benefits and future increases in the real price of delivered energy are thus ignored. Furthermore, no attempt is made to place a conservation value on diminishing supplies of fossil fuels.

The potential for energy saving in buildings and the conservation of particular forms of energy would increase significantly if these aspects were taken into account. That potential can be achieved by means of the ECIY approach.

The approach is based upon an existing and well-established approach, known as life cycle costing, but rearranges the component parts. Design teams express their proposals in terms of energy conservation investment factors (ECIFs) and clients express their investment criteria in terms of energy conservation investment yardsticks (ECIYs). Both are expressed in kW h/annum/£NPV spent. The acceptability of a proposal is determined by a simple comparison of its ECIF with the client's ECIY.

In the public sector, the ECIY and the relative value of each fuel type can be prescribed by the Government. In a similar manner, incentive schemes and guidance material could be introduced into the private sector. The relative values prescribed for each fuel type also provide a useful design tool to help designers consider alternative methods of saving energy. The present values assumed for each fuel type can also be adopted in determining fuel policies for buildings.

Although complex in theory, the approach is simple to use. As an administrative tool, it has the potential to enhance the effectiveness of national policy.

PART 2

The Contribution of Solar Water Heating to Low Energy Hospital Design

This part of the chapter describes the section of the DHSS Low Energy Hospital Study covering the potential energy contribution that solar water heating can have to reducing energy consumption in new hospitals. It highlights some of the analysis and studies carried out and examines the integration of solar systems into new hospitals. The importance of the need to discriminate between the energy performance of individual stand-alone solar water heating systems as compared to systems integrated with whole hospital energy strategies which include recovery arrangements is stressed. When viewed in this way the energy contributions of solar water systems can be considerably diminished when competing with recovered heat sources, to the extent that their incorporation becomes more difficult to justify when the economic considerations are taken into account.

Over the last ten years or so, due to the increase in cost of fuels, considerable interest has been shown in the application of solar energy. This interest has fostered the growth in the UK of a small but significant industry in manufacturing and installing solar systems and has stimulated considerable research development and demonstration activity.

The characteristics of solar energy as an energy source can be considered to cover three primary forms of application: photochemical, photoelectric and thermal. In this part of the chapter attention is focused on thermal applications and more particularly solar domestic hot water heating applied in the context of low energy hospital design.

For thermal uses of solar energy, the temperature of use is most important. High temperatures, for example for power generation require concentration of the sun's rays and complex tracking mechanisms. Intermediate temperatures for process heat or absorption cooling required low concentration collection or sophisticated stationary collectors. Whereas domestic hot water heating needs only relatively simple flat plate collectors. Space heating can be achieved either with flat plate collectors or by the building elements themselves collecting the incident radiation. This latter approach, often referred to as passive solar heating, is considered in terms of the building's thermal performance.

An active solar system, as distinct from a passive solar system is one which uses a collection device to convert the incident radiation to heat, transferring this to a heat transmission medium (usually water), which in turn is distributed to the point(s) of utilization. The system will often include other elements such as thermal storage, auxilliary heat sources and controls. The most common application of active solar systems in the UK and elsewhere is in domestic hot water service heating.

UK research, development and demonstration of solar water heating technology has been undertaken or commissioned by a number of organizations, including universities (notably the Solar Energy Unit, University College, Cardiff), the Building Research Establishment and the Energy Technology Support Unit, acting for the Department of Energy. Field trials for both domestic and non-domestic applications include, for example, a major installation for a catering facility at Torbay Hospital in the South Western Region.

Close liaison is maintained between the research community, government bodies and the commercial and industrial sectors. The UK has also had involvement with European and International Solar R&D programmes and the UK section of the International Solar Energy Society provides an important additional focus for ideas and information. All this has been instrumental in improving equipment design and system performance and information.

Guides to good practice are to be found in the British Standard Code of Practice [5] and the HVCA guide [6]. The British Standard includes descriptions of system types, component and system design considerations and a widely accepted performance prediction method. The HVCA guide provides useful and straightforward engineering advice, including a section on cost effectiveness. Another reference which complements the above two guides is the BRE Report *Solar Heating Systems for the UK: Design, Installation and Economic Aspects* [7].

ENERGY DEMAND PATTERNS AND GRADES

Seasonal variation of incident radiation limits potential applications since for space and air heating systems the maximum demands occur in the winter when the available solar energy is at a minimum. Interseasonal storage could compensate for this, but this option is not considered economically viable at the present time. The demand for domestic hot water, however, is maintained at a uniform level throughout the year and for a hospital occurs mainly during the daytime. Solar domestic hot water heating is, therefore, a more attractive proposition.

The 'grade' of heat, or temperature required, has a direct bearing on system efficiency as the temperature achieved by a collector is that at which a balance exists between the radiation received and the combination of usefully transferred heat and heat losses. This aspect is covered later when considering collector types, but the general rule is that heat losses increase proportionately with increased temperature. Therefore, for a given level of incident radiation, higher temperatures result in less heat being available for use. Another way of viewing this is that, for a given required temperature, a threshold level of radiation must be available to offset the heat losses. Only when the intensity of radiation is above this level can any heat be usefully collected.

SOLAR COLLECTORS

A range of collector types are available. Performance varies considerably but physically similar types can be grouped together under three main categories:

(a) Flat plate collectors.
(b) Evacuated tube collectors.
(c) Concentrating collectors.

The flat plate collector is by far the most common type and consists of a blackened or treated absorber plate. This plate converts the incident radiation to heat which is transferred to a fluid heat transmission medium flowing through passages formed in or bonded to the plate. The fluid is usually water (sometimes containing an antifreeze solution), but can also be air or a low volatile mineral oil. Except for low temperature application, such as for swimming pools, the absorber plate is backed by insulation and fitted within a casing with one or more transparent covers. The purpose of the transparent cover is to reduce heat losses due to both convection and long wave radiation. A second transparent cover, although it can significantly reduce convection heat loss, does little to further reduce the radiation loss in a climate with a preponderance of diffuse radiation. Its beneficial effect can be more than offset by the reduction in energy transmitted through the covers due to reflection.

Improved performance can be obtained by the use of special coatings on the collector plate. These selective surface treatments provide high absorptivity to incoming short wavelength radiation and low emissivity of outgoing long wavelength radiation. A coating operating on a similar principle is also available for the transparent cover. In this instance, applied to the inward facing surface it is selectively transparent to incoming short wavelength radiation and highly reflective to the outgoing long wavelength radiation.

DESIGN CONSIDERATIONS

Key factors to be determined in relation to solar system design and which have a major bearing on performance are:

(a) Availability of solar energy.
(b) Pattern of anticipated use.
(c) Grade of heat or temperatures required.
(d) Efficiency of collection.
(e) System configuration.

The most significant aspect of solar energy availability is its variability both diurnally and seasonally. Based on data from Kew a horizontal surface will receive about 900 kW h/m² of solar radiation a year but the amount received in mid-summer is about ten times that received in mid-winter. Geographical variation within the UK is not very great although there is a marked difference between northern and southern locations in the winter months. Another important aspect is the proportion of diffuse radiation. The maritime climate of the UK results in a high incidence of cloudy conditions particularly in winter months. Due to this over 50% of total incident radiation is diffuse, although this also is subject to regional variation.

The quantity of solar energy intercepted by a collector is affected by the tilt angle and orientation of the surface, but due to the high proportion of diffuse radiation this is not as crucial an issue as might first be thought. This is well illustrated by Fig. 2.1 (Chapter 2) reproduced here for convenience as Fig. 5.7. This shows for a reference domestic hot water installation the effect of a departure

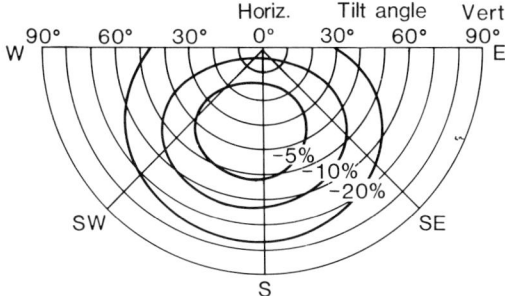

Figure 5.7. Variation of the solar energy supplied by the reference system with a class IV collector, with orientation and tilt of the collector (reproduced from BS 5918 by permission of the British Standards Institution)

from optimum tilt and orientation. The optimum is seen to be at 45° tilt just slightly to the west of due south, but for a collector facing anywhere between south-east and south-west tilted between 5° and 60° from the horizontal the reduction in annual performance is at most only 15%. An increase in tilt angle will improve solar collection during the winter but this is generally to the detriment of overall annual performance. Shallow tilt angles may affect performance due to the accumulation of dirt. Problems of rain-noise and hail-breakage also become more acute.

Evacuated tube collectors are a relatively recent development. Only one UK firm is so far involved in the manufacture of this type and not yet at the scale of mass production. The collector consists of strips of selectively coated absorber surface contained within evacuated glass tubes. Thermal losses are thereby about one third of those for a single glazed, non-selective flat plate collector. Other advantages are that the absorbers in individual tubes rather than the entire collector can be tilted at the required angle giving much greater freedom in location. They are therefore expected to be very durable as the selective absorber surface is in a vacuum and is unlikely to degrade. Degradation has sometimes proved to be a problem when these surfaces are exposed to the chemicals and moisture present in the atmosphere.

Some manufacturers of evacuated tube collectors employ heat pipes as the means of extracting the heat. These enable a single metal-to-glass seal thereby reducing problems of differential thermal expansion, and facilitate the removal and replacement of individual units within a collector array. Because of its gravity dependence, however, the units need to be arranged so that the heat pipe slopes at an angle of about 20° to the horizontal.

The collectors mentioned so far are usually fixed in a single position. Concentrating collectors, however, need to follow accurately the path of the sun in order to keep the radiation focused on the absorbing surface. It is only possible to focus the direct component of the available incident radiation and various means are employed to achieve this. These include Fresnel lenses, parabolic trough reflectors, or an array of plane reflectors directing the radiation to a tower-mounted collector. Concentration ratios can be very high. The absorbing surface is, therefore, compact and heat losses are relatively small. Where the climate provides a high proportion of direct radiation, such as in arid regions, very high temperatures can be obtained at high efficiency. In the UK climate however, the low proportion of direct radiation imposes a severe limitation.

SOLAR WATER HEATING SYSTEMS

There are many possible system configurations for solar water heating. Systems can, however, be classified according to certain features. For example, a system may be either direct or indirect, depending on whether the hot water service water passes through the collector panels. The system may also have either a gravity (thermosyphon) or pumped circulation and the circulation can be either open vented or sealed, that is fitted with a pressurization/expansion unit. A gravity circulation system has the advantage of simplicity as it requires neither pump nor controls, but storage has to be incorporated above the level of the collectors which can be difficult to arrange.

Freeze protection is an important consideration and can be achieved by the addition of antifreeze or the use of a mineral oil heat transfer fluid or by arranging for the system to be drained down when freezing conditions are likely. The former two methods are only applicable to indirect systems and the drain-down, although applicable to both, should be questioned in direct systems as it may accelerate the deterioration of circuit components unless great care is taken in material selection and installation. With sealed indirect systems, at the risk of greater complexity, a drain-down arrangement could be considered which charges the collectors with nitrogen whilst drained.

For hospital application solar water heating systems could be either integrated into the low or medium grade heat circuits in parallel with heat recovery systems, or linked into the domestic hot water service as a dedicated system. In the latter instance the requirement that domestic hot water should not be stored at temperatures below 60°C for bacteriological reasons should be borne in mind.

Low temperature applications allow solar collectors to operate at higher efficiencies. Integration with the low grade heat circuits would, therefore, be more advantageous than with the medium grade systems. Preheating of the domestic hot water service from mains water temperatures provides ideal conditions for an effective solar system, providing that a satisfactory solution can be found for the storage temperature problem.

Another factor which limits the utilization of solar energy is the magnitude of the heat demand. This is also the case with heat recovery systems but affects solar systems more acutely due again to the variable nature of solar radiation. If a solar system is designed to meet a given demand at maximum conditions of incident radiation, the annual contribution will be very low. If designed to meet the demand at more moderate conditions, the annual contribution will be increased relatively but more heat than can be utilized will be available for some of the time. Increasing the size of the collector array with a given demand therefore results in a diminishing contribution per unit of surface area. This again becomes primarily an economic issue.

Storage provides a means to resolve mismatches between the time of energy availability and of demand. Diurnal and day-to-day variations in the availability of solar energy make storage an important issue for solar systems. If the pattern of heat demand is fairly continuous and the scale of the solar contribution is small, then storage could be omitted. For domestic hot water systems, however, the diurnal variation in demand is fairly marked, therefore storage is necessary if a worthwhile solar contribution is to be obtained.

Two possible methods of incorporating thermal storage are illustrated in Figs 5.8 and 5.9. The system shown in Fig. 5.8 uses the conventional arrangement of storage cylinders and in this case the solar system controls are set to allow

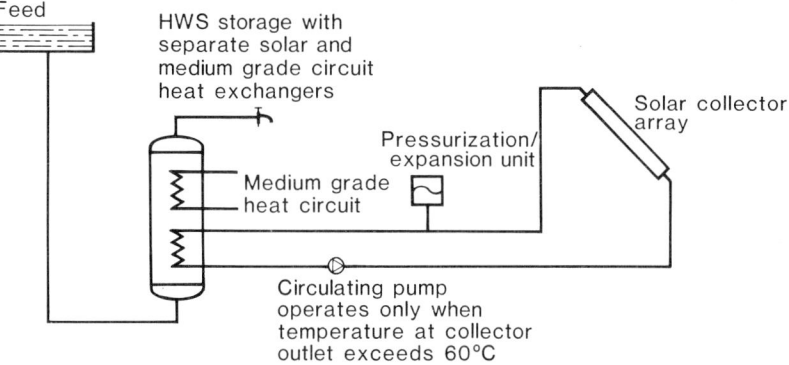

Figure 5.8. Solar domestic hot water heating – arrangement using medium grade solar heat only

circulation between the storage cylinders and the collector array only when the flow temperature from the collectors is above 60°C. This arrangement operates as a medium grade heat circuit and with minor modification could be integrated as a component part of a centralized medium grade heat system. The bacteriological problem is overcome, but at the expense of collection efficiency.

Figure 5.9 shows an alternative arrangement in which the thermal storage is incorporated as part of the primary circuit. A plate heat exchanger is used to transfer heat from the primary circuit to the mains water inlet of the domestic hot water system. For much of the year the temperature of the water leaving the plate heat exchanger will be between 30° and 45°C, but this does not present a problem as the residence time before reheat in the conventional storage calorifier is short. With this arrangement a drain-down system of frost protection is indicated.

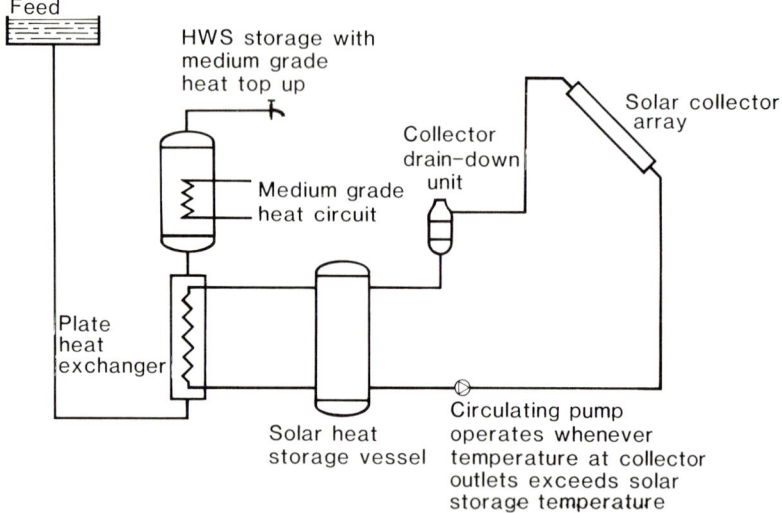

Figure 5.9. Solar domestic hot water – preheat arrangement

Alternatively the thermal storage vessel could be separated from the collector circuit by another plate heat exchanger, the smaller volume of the collector circuit allowing antifreeze solutions or mineral oil transfer fluid to be considered. The addition of another plate heat exchanger would, however, impose a further reduction in system effectiveness.

The amount of thermal storage to be provided is primarily an economic issue. There is a relationship between storage quantity, load demand pattern and collection efficiency. If the storage volume is small the extent to which supply and demand patterns can be matched is limited. The storage temperature will also be relatively high, which reduces collection efficiency. Conversely, if the storage volume is large, collection efficiency will be raised, but the mean temperature of storage will be lower. For a given demand temperature requirement this limits the solar contribution. However, studies undertaken at the Solar Energy Unit, University College, Cardiff have demonstrated that above a certain minimum value solar domestic hot water preheat systems are fairly insensitive to storage quantity. Optimum storage quantities for domestic scale systems, for example, have been found to lie between 35–120 litres per square metre of collector area. In larger systems the pattern of demand is less variable, which would tend to extend the range of optimum conditions to lower volumes.

Active solar systems present designers with certain constraints which may be difficult to resolve. These include:

(a) The visual appearance of potentially large areas of collector surface and the response to this of the planning authorities.
(b) The methods by which the collectors are fixed or integrated with the building

fabric, and the associated problems of windloading, weathertightness, structural stability, condensation and ventilation.
(c) Provision of access for maintenance.

There may also be detailed issues which will require consultation with regulatory authorities, such as the National Water Council and the Fire Officer. These often result from requirements which pre-date the current interest in solar technology.

COLLECTOR PERFORMANCE AND CONTRIBUTION

In order to determine the potential solar contribution in the context of the hospital a study on solar collector performance was commissioned from Solar Energy Developments (SED).

The first part of this study was concerned with the performance of a range of solar collectors at specified operating conditions considered to encompass all likely water heating applications. Collector types analysed ranged from a single glazed matt black flat plate to a tracking concentrator. Operating conditions included preheat of domestic hot water from mains water at 10°C through to a medium grade heat circuit with a flow temperature of 65°C. At this stage other aspects of the solar systems, such as size of collector array, storage volume and load pattern were not specified. The results imply ideal conditions where the full potential is utilized and are, therefore, only indicative of the relative performance of the various collector types.

Six collector types were selected for analysis. These were:

(a) Single glazed matt black flat panel.
(b) Double glazed matt black flat panel.
(c) Single glazed selective surface flat panel.
(d) Single glazed selective surface air collector.
(e) Evacuated tube selective surface.
(f) Tracking/parabolic trough concentrator.

The performance of these collectors can be characterized by two parameters: optical efficiency and thermal losses per unit area. The values taken for these parameters for each of the collector types are given in Table 5.3 and are also shown in Fig. 5.10 superimposed on an illustration taken from BS 5918 [5] which gives typical zones of values for a number of collector types. Some values are seen to deviate slightly from the BS 5918 zones as allowance was made for more recently available test data. Values for the tracking/concentrator type assume a collector with a concentration ratio of 20.

Eleven operating conditions were specified although it was not considered necessary to analyse the performance of all six collector types at all of these conditions once a trend in the results became apparent. The specified conditions

Table 5.3 Solar collector performance parameters. Values of optical efficiency and heat loss rate

Collector type		Optical efficiency	Heat loss (W/m² °C)
(a)	Single glazed matt black	0.75	6.0
(b)	Double glazed matt black	0.65	4.5
(c)	Single glazed selective surface	0.75	4.5
(d)	Single glazed selective air collector	0.70	6.0
(e)	Evacuated tube selective surface	0.70	2.0
(f)	Tracking/concentrator	0.65	0.75

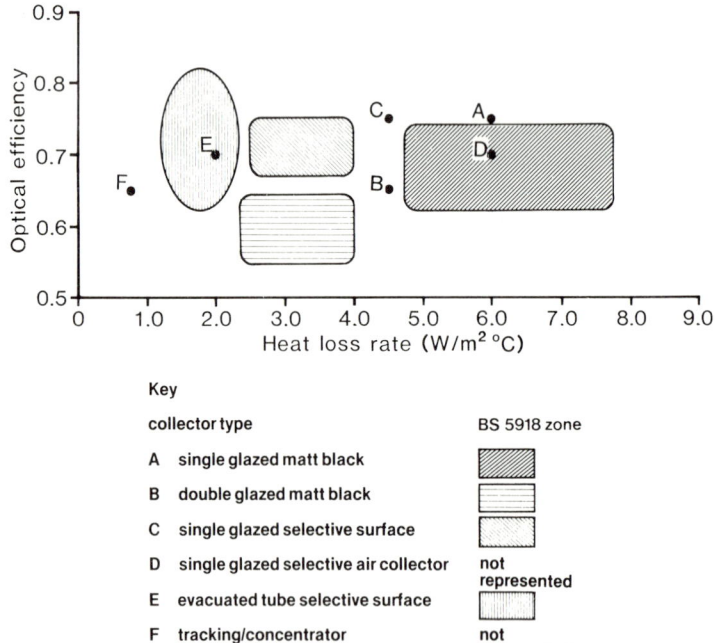

Figure 5.10. Collector performance parameters. Values of optical efficiency and heat loss rate compared to approximate zones of values given in BS 5918

were as given in Table 5.4. Conditions 9, 10 and 11 are for a fixed water flow rate and variable return temperature.

Calculations for the non-tracking collectors were carried out using a method developed by Klein [8]. The method employs average monthly radiation and ambient temperature data which, for consistency with other climate data used in the study, were taken from Kew 1967. Collectors were taken to be orientated due south and inclined 45° from the horizontal.

Table 5.4 Specified solar collector operating conditions

Condition No.	Inlet temperature (°C)	Outlet temperature (°C)	Flow rate (kg/m² s)
1	25	35	variable
2	40	50	variable
3	55	65	variable
4	10	35	variable
5	10	50	variable
6	10	65	variable
7	2	50	variable
8	25	65	variable
9	25	variable	0.015
10	40	variable	0.015
11	55	variable	0.015

The tracking/concentrating collector could not be analysed by the same method as the non-tracking types as only direct radiation can be used and the 'cosine losses' suffered by fixed collectors are avoided. A method of analysis was developed in which average monthly efficiencies were derived from hourly data, this average efficiency then being applied to average monthly incident radiation. For the hourly values the effect of overshadowing from adjacent collectors was calculated assuming a 'polar mounting' with the parabolic trough parallel to the earth's axis and a space between collectors equal to their width.

Results of the analysis of both tracking and non-tracking collector types are given in Table 5.5 in terms of the annual energy supplied per unit area of collector for each of the specified operating conditions. The fall-off in collector performance at high temperatures is apparent, particularly for the single glazed matt black and air collectors. Conversely the tracking collector only shows an advantage over other types at high temperatures. The selective surface single glazed collector is seen to perform well under all operating conditions and the potential for evacuated tube collectors when they become more generally available is very clear. Finally, a general indication of the merits of preheat systems for domestic hot water service application as compared to medium grade heat input can be gauged by comparing the results for conditions 3 and 4 across all collector types.

HOSPITAL APPLICATION AND ENERGY CONTRIBUTION

Following on from the above comparative study, the performance of a solar domestic hot water service preheat system of the type shown in Fig. 5.9 was analysed. The objective of this part of the study was to examine the effect the size

Table 5.5 Comparative performance of solar collector types

					Performance at specified operating conditions (kW h/m² per annum)						
	1	2	3	4	5	6	7	8	9	10	11
(a) Single glazed matt black	403	250	151	489	350	231	305	214	455	285	175
(b) Double glazed matt black	376	249	162	447	340	244	298	224	418	279	183
(c) Single glazed selective surface	461	323	224	538	430	330	379	300	497	349	243
(d) Single glazed selective air collector	362	217	126	443	–	–	–	–	–	–	–
(e) Evacuated tube	532	451	381	577	–	–	–	–	–	–	–
(f) Tracking/concentrator	346	333	312	362	–	–	–	–	–	–	–

of the collector array has on the solar contribution to the overall domestic hot water service energy requirement.

For the basis of the study a typical nucleus district general hospital of 300 beds was used as the datum for the application of solar water heating [9]. The content and general arrangement of the hospital is shown in Figs. 5.11 and 5.12.

Five collector array sizes were considered, with areas of 350, 500, 1000, 1500 and 2000 m². The daily domestic hot water service demand was arranged as 48 000 litres to be raised from a mains water inlet temperature of 10°C to a final storage temperature of 60°C. Calculations were carried out using a method developed by Brinkworth [10]. This method, as for the Klein method [8] employed earlier, uses average monthly radiation and ambient temperature data.

The results, given in Table 5.6 are for single glazed selective surface collectors, as this type emerges from the earlier study as having a good performance over the

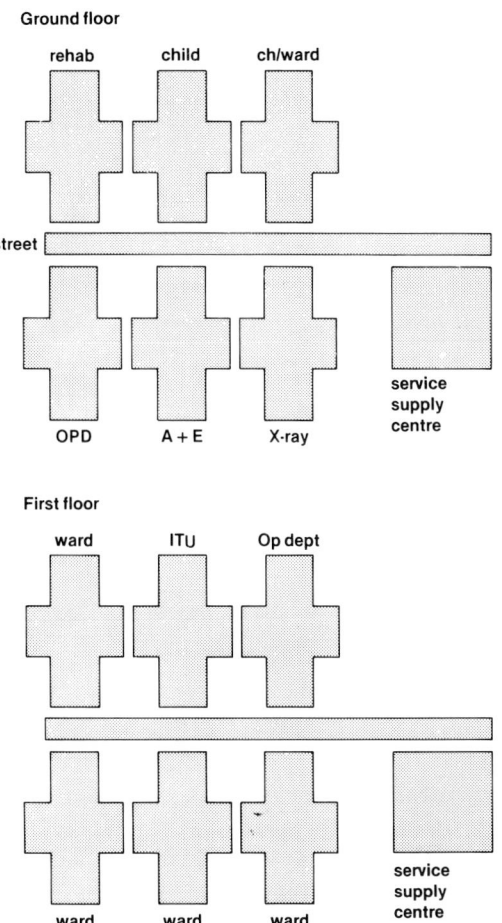

Figure 5.11. Typical content of nucleus hospital

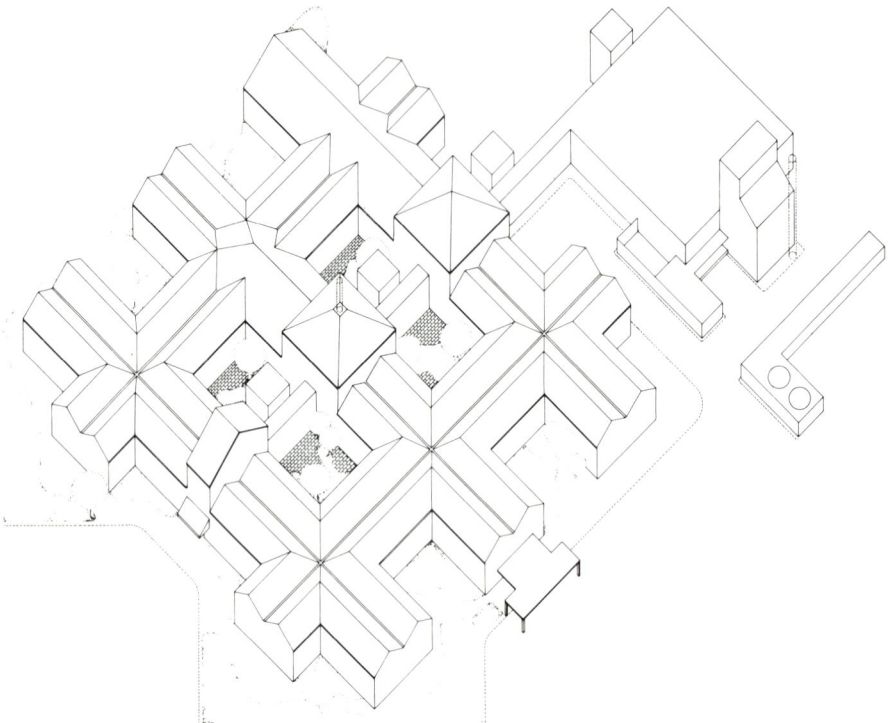

Figure 5.12. General arrangement of building forms

Table 5.6 Solar domestic hot water preheat system. Influence of collector area

Collector area (m²)	Solar contribution (kW h/annum)	Proportion of net domestic hot water service heat load
350	182 000	0.16
500	235 000	0.20
1000	361 000	0.31
1500	444 000	0.39
2000	505 000	0.44

anticipated temperature range. As expected, increasing the collector area increases the annual solar contribution, but not in direct proportion. At 350 m² the solar contribution amounts to 519 kW h/m² whereas at 2000 m² this falls to 252 kW h/m². Evacuated tube collectors are not yet readily available in the UK but for purposes of comparison calculations were made using this type for two array sizes, 350 and 1000 m². The resulting annual contributions were 548 and

$445\ kW\ h/m^2$ respectively, which would seem to indicate that the decline in performance with increasing collector area is less marked than for the single glazed selective surface collector.

Although the available roof area of the type of hospital considered could accommodate $2000\ m^2$ of solar collectors, the arrangement would necessarily have to be dispersed and fragmented thereby increasing the size and complexity of heat distribution and associated losses. A more compact arrangement is possible on the roof of the service/agency centre which is ideal as distribution distances are minimized. To maintain appropriate orientation and tilt angles, collectors need to be spaced so as to minimize overshadowing by adjacent collectors. Total collector area is, therefore, generally less than half of the available roof space. On this basis a collector array of $500\ m^2$ is the maximum that could be accommodated at this location.

A solar system for preheating the domestic hot water service with a collector array of $500\ m^2$ represents a reasonable compromise between collection efficiency and overall annual energy contribution. Figure 5.13 shows the contribution that such a system would make in reducing the prime energy requirement of the domestic hot water service system. The nett contribution of $235\,000\ kW\ h/annum$ results in an overall saving of $328\,000\ kW\ h$ when the reduction of primary circuit distribution and generation losses are taken into account. There is, however, a small increase in electrical energy used for circulating pumps assessed at $2000\ kW\ h/annum$. Figure 5.14 presents the data in the form of annual profiles before and after incorporation of the solar system.

The solar contribution is shown without discriminating between low (30–40°C) and medium grade heat (70–80°C). During the winter period the energy saved will be almost entirely low grade, whereas during the summer period a significant medium grade contribution can be expected. It is necessary to make this distinction when considering the integration of a solar system into an overall energy strategy. This is particularly important when the potential availability of low grade heat from recovery systems is considered.

The dramatic effect that this can have in limiting the effective contribution is an issue which must be considered in a total hospital context. The DHSS Low Energy Study has highlighted the energy benefits of incorporating into the future generation of low energy hospitals an energy strategy which allows for recovered heat from sources such as incineration exhaust air and refrigeration.

Furthermore, it has discerned that the recovery strategy should be arranged so that recovery systems have the ability to transfer heat within and across the boundaries of the individual systems to which they are applied. For example, recovered heat from exhaust air should be able to be used in any system – air treatment, space heating, domestic hot water – which has a deficit of low grade heat.

Adopting this approach has considerable benefits, particularly in winter and midseason when the recovered heat can reduce the prime energy heating input by amounts approaching 50–60%. In the summer periods the recovered heat is for

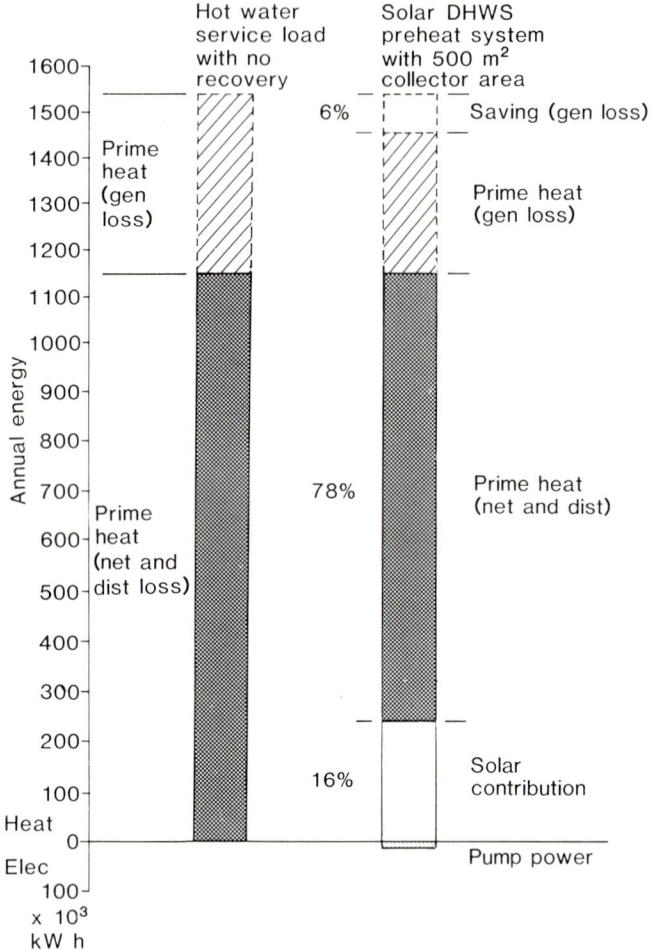

Figure 5.13. Solar domestic hot water preheat – performance profile

much of the season more than adequate to satisfy all the heating associated with heating and hot water service preheat. Therefore, the potential and contribution of solar heat for this purpose is to a very large extent largely diminished. It is this aspect which has an important bearing on the effective incorporation of solar water heating into new hospital projects.

ECONOMIC AND FINANCIAL ASPECTS

Whilst it is not possible to provide all the financial and economic studies which were carried out as part of the DHSS study, some comment and information is

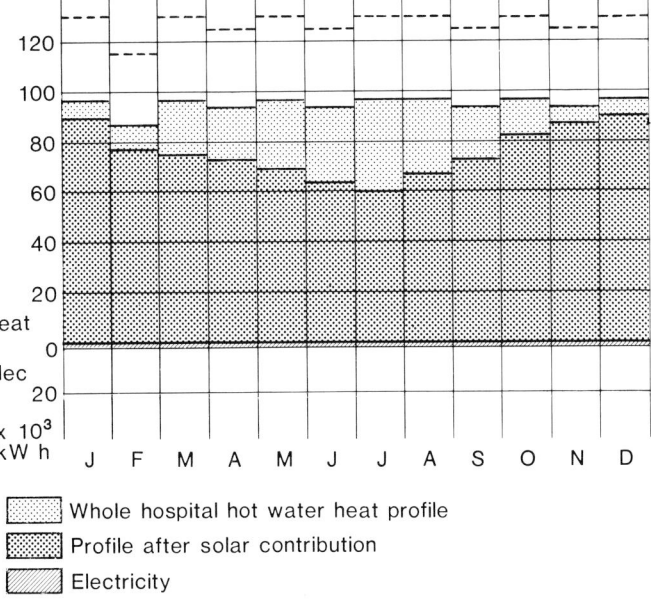

Figure 5.14. Solar domestic hot water preheat – contribution towards whole hospital hot water heat requirement

included to highlight the impact and energy contribution of solar water heating when viewed in a whole hospital context.

For an individual stand-alone solar water heating system the energy and financial implications obtained from the study were as follows. These are based on a low grade solar water heating system applied to the domestic hot water system and comprising some 500 m² of flat plate collector panel area.

(a) Capital and non-fuel revenue costs amount to some £NPV 187 000.
(b) Unfactorized units of energy for fossil fuel amount to a saving of 328 000 kW h/annum and an increase in electrical energy of 2000 kW h/annum.
(c) Factorized units of energy and ECIF were as follows, based on fuel weighting factors of 1.71 for gas, 1.90 for oil and 5.84 for electricity:

    Gas/electric mix   549 000 kW h/annum
         ECIF   2.94

    Oil/electric mix   611 000 kW h/annum
         ECIF   3.27

For a solar water heating installation forming part of an overall integrated scheme involving a number of energy saving measures including recovery from incineration and exhaust air the energy and financial impact were as follows:

(a) Capital and non-fuel revenue costs amount to £NPV 187 000, for the addition into the scheme of solar water heating.
(b) Factorized units of energy and ECIF were as follows, based on fuel weighting factors previously stated:

Gas/electric mix   120 000 kW h/annum
ECIF   0.49 reduction

Oil/electric mix   134 000 kW h/annum
ECIF   0.52 reduction

It should be stressed that the energy and ECIF values are the change that takes place as a result of the incorporation of solar water heating into a whole hospital scheme which already incorporates a range of energy saving measures including various recovery systems and is based on adopting a totally integrated approach.

It can be seen that when solar heating is applied in a whole hospital context that the energy contribution is reduced compared to its value when viewed as a stand-alone measure. Furthermore, the effect on the overall ECIF for the cumulative effect of all the energy measures is lowered. For comparative purposes the overall ECIF for a low energy hospital, without solar water heating but achieving energy reduction of 50%, is in the region of 6.0.

CONCLUDING COMMENTS

Clearly the energy benefits and financial implications of incorporating solar water heating into new hospitals are dependent, to a large degree, on the overall energy strategy adopted for the whole hospital. All too often measures have been assessed on their stand-alone performance and as has been shown this can vary significantly, when viewed in combination with other measures.

The findings of the DHSS Low Energy Study show conclusively that the incorporation of solar water heating is difficult to justify on energy investment grounds, especially when energy savings approaching 50% and above can be achieved with other more effective measures in terms of investment and energy contribution.

The large amounts of recovered energy available from sources such as exhaust air and incineration if properly arranged can offset, to a large degree, the low grade heating requirements of the hospital during midseason and summer periods. In so doing this diminishes the effective contribution that solar water heating can make in low energy hospitals.

However, it should be stressed that changing technology and manufacturing techniques of solar panels may lead to more effective cost equations and that this in turn could influence the ranking of solar water heating in the energy measures being considered and therefore result in its inclusion in future projects. The accounting procedure explained in the first part of this chapter provides a basis for this evaluation.

ACKNOWLEDGEMENTS

We acknowledge Solar Energy Developments for work as specialist consultants for part of the solar analysis studies and also the Solar Energy Unit, University College, Cardiff.

REFERENCES

1. Department of Energy, *Energy Trends*, monthly bulletin.
2. HM Treasury Public Sector Economic Group (1980) *Investment Appraisal and Discounting Techniques and the Use of the Test Discount Rate in the Public Sector*, HMSO, London.
3. Department of Energy (1979) *Energy Projections*, HMSO, London.
4. Building Research Establishment (1975) *Energy Conservation: A Study of Energy Consumption in Buildings and Possible Means of Energy Saving in Housing*, Working Party Report, CP56/75, BRE, Garston, Herts, UK.
5. British Standards Institution (1980) *Solar Heating Systems for Domestic Hot Water*, BS 5918, BSI, London.
6. Heating and Ventilating Contractors Association (1979) *HCVA Guide to Good Practice. Solar Heating for Domestic Hot Water*, HVCA, London.
7. Building Research Establishment (1979) Solar Heating Systems for the UK. Design Installation and Economic Aspects, BRE, Garston, UK.
8. Klein, S. A. (1978) Calculation of flat plate collector utilisability. *Solar Energy*, **21**, 393–402.
9. Department of Health and Social Security, *DHSS Nucleus Hospitals*, Various technical studies and design data, DHSS, London.
10. Brinkworth, B. J. (1979) Asymptotic behaviour as a guide to long term performance of solar water heating systems. *Solar Energy*, **21**, 269–282.

DISCUSSION

Dr S. J. Wozniak (Building Research Establishment). In Part 2 of the chapter a 50 to 60% saving of energy has been quoted. Is this in primary or delivered terms? Reference has also been made to heat recovery from exhaust air and other low grade heat systems. Coefficient of performance is a crucial factor and it would be interesting to hear what values have been taken. A third question relates to the efficiencies of the solar collectors. From the figures presented these vary, although outputs as high as 400 to 500 kW h/m² year have been quoted. What collector types do these higher values apply to?

J. R. J. Ellis (Building Design Partnership). The overall savings quoted are in delivered site energy terms and not primary energy. The difficulty of relating to primary energy is that the factors which are applied to convert the primary energy are extremely contentious depending on what parameters and view is taken on the off-site values which extend far beyond just those often published as being, so called, representative of generation and distribution losses.

The accounting method described in Part 1 of the chapter does allow for conversion to a primary basis and this is achieved by applying weighting factors to the various fuel types. These can be adjusted and applied at whatever values are considered appropriate for the circumstances of the account and can be related to such aspects as fuel cost or price, generation and distribution efficiencies, national or local energy policies, etc. The method provides a very flexible approach to this aspect of energy accounting.

On the question of heat pump performance the study showed that systems which incorporate both natural and heat pumping thermal exchange were favourable to those just relying on one type of exchange process. Systems were developed which are capable of centralized recovery of low grade heat (30–40°C) with common store facilities which can supply any system – air treatment, space heating, domestic hot water – having a low grade heat deficit.

Our studies on heat pump performance showed that coefficients of performance of 5.0 and higher were achievable under certain operating conditions. I should stress that this is the COP of the heat pump and relates to the recovered heat at the condenser compared to that required at the inlet shaft.

The question on the efficiency of the solar collectors could perhaps best be answered by John Field of Solar Energy Developments. SED were appointed by BDP to carry out some detailed simulation and analysis work on comparative panel performance and solar contribution.

John Field (Solar Energy Developments). As described in the chapter a wide range of solar collector types and operating conditions were explored and only the summarized results are presented here. Tables 5.3, 5.4 and 5.5 show the data results for the comparative studies which are based on ideal conditions where the full potential of the collector is assumed. Various system simulation studies were carried out and the results of the single glazed selective surface collectors are shown in Table 5.4. These are based on applying solar water heating of varying collector area to a hospital water system with a demand of 48 000 l/day. Results show that for a collector area of 350 m² the annual contribution amounts to 519 kW h/m². At 2000 m² the figure reduces to 252 kW h/m². Using a collector area of 350 m² results in a high figure because of the relatively small system size and the large domestic hot water demand characteristic. However, at 350 m² of collector area the proportion of net domestic hot water load contributed by solar is about 16%.

J. R. J. Ellis. I should just comment on the solar conditions for the water side variants shown in Table 5.4. These were selected at a time when the precise arrangement of the solar system was unknown. Various options were being considered at that stage and these included integration with the low or medium grade heat circuits as well as employing a simple dedicated preheat system to the domestic hot water system only.

A. R. Tanner (Wessex Regional Health Authority). Reference was made in Part 2 of the chapter to various grades of heat being identified in the study. Could this comment be expanded and some indication given to the grades and temperature ranges.

J. R. J. Ellis. When analysing the system heat requirements for the hospital it was important to classify these in terms of their temperature grades. Three heat grades were established, namely high, medium and low. Taking these in order high grade was established as 100°C or higher and represented those systems where steam or medium temperature hot water was required, such as for humidification and sterilizing. Medium grade represented a temperature range of 70–80°C and low grade 30–40°C. Much of the environmental space heating and air treatment and the process hot water service can be satisfied essentially with low grade heat, which corresponds to the temperature range from many of the recovery heat sources.

T. J. Wyatt (Brown Crozier & Wyatt). Does Mr Ellis find that by increasing window areas by from 25–35% annual energy savings on lighting are actually obtained and is this also true for the overall energy requirement of the building?

J. R. J. Ellis. The short answer is that we do believe the savings can be made. The glazing areas were increased to provide the required levels of daylight and optimized on both the use of natural lighting and passive solar – both of these are important and inter-related. The different switching patterns were examined but the main aim is to create a quality environment for the wards. The distribution pattern of natural lighting is good and we have concentrated on using the natural lighting to produce a good environment qualitatively and in this way dissuade people from wanting to switch on the artificial lights. If some of the artificial lighting is task lighting this too can give benefits.

J. Campbell (Ove Arup Partnership). It is important to have a switching arrangement that enables people in the building to take advantage of the savings by choosing zones related to the availability of natural lighting.

A. R. Tanner. I was very interested in the life-cycle costing techniques described. Could their relative merits be enlarged upon in relation to simple present worth techniques? As a designer, I always use simple present worth techniques including sensitivity analysis. What are the additional advantages of the techniques put forward?

D. Allen (Building Design Partnership). The 'life-cycle costing' technique seeks to compare the costs and benefits of a proposal over time. It does that by 'discounting' the costs and benefits from when they occur back to a common time datum and expressing them in terms of their 'present-worth'. Presumably, this is the 'simple present-worth' technique to which you refer.

In applying this technique, one has to distinguish between costs and benefits. Traditionally, 'benefits' may be viewed as reducing revenue costs and 'costs' as the extra capital cost expended in achieving such reductions. The objective of the study was to reduce energy consumption. Reductions in energy consumption, therefore, are the 'benefits' being sought by the proposals, and the 'costs' are the capital and other revenue implications of achieving those energy reductions. Again, traditionally both the 'costs' and the 'benefits' are normally expressed in financial terms.

Because of the relatively static nature of the 'costs' and the relatively volatile nature of the 'benefits' (due to the relative effect of inflation on fuel prices) over time, the particular approach described expresses the 'benefits' in units of energy. Expressed in this way, the guidance material is presented in a more neutral form and the sensitivity of a proposal to different fuel-types and assumed price scenarios is readily apparent. It also enables investment criteria to be considered separately and sensitively by clients. Furthermore, it enables energy conservation to be controlled by factors other than price.

R. Burton (Taylor Young & Partners). Has the project team considered the possibility of glazing over the courtyards to form an atrium. Walls of lighter weight and a cheaper form of construction could be utilized but total heat losses through external walls would be reduced as a consequence.

A US hospital is designed in this way and is successful. Ventilation of the rooms facing the atrium is achieved by mechanical pressurization of the atrium itself.

J. R. J. Ellis. It was considered but rejected because the walls are well insulated and account for a very low percentage of heat loss. The main problem is that many of the perimeter spaces, particularly the wards and the consulting rooms which form a large percentage of the perimeter, rely on opening windows for losing excess heat by natural ventilation in summer. This brings the complication of possible cross-infection by movement of air from department to department via closed-over courts, even though they may be naturally ventilated atria at times. The Philadelphia Hospital is in a different climate and I suspect many of the perimeter rooms are mechanically ventilated or air conditioned.

J. Keable (HELIX Multi Professional Services). It has been explained that a whole range from 350 to 2000 m² of solar collector areas have been examined. Our own studies have also centred on optimization of collector area and to look at differing collector/system arrangements as clearly there becomes a point at which increasing the scale of the installation begins to show diminishing returns. Overlaid on this, of course, is the efficiency of the collector unit itself. Whilst both these aspects have been highlighted as part of the study it occurs to me that any significant trend could affect the growth and direction of the solar collector manufacturing industry and hence the resultant price of equipment.

J. R. J. Ellis. Obviously solar technology and equipment development is in its infancy in the UK, particularly in its application to the non-domestic market. Practical feedback on the operation, maintenance and performance of large scale installations is limited, although the South Western Regional Health Authority and the Department of Energy are installing a major prototype installation at Torbay Hospital. The work of the DHSS study has determined the potential contribution that solar energy could make to a 300-bed nucleus hospital. Consideration of the performance data, technology considerations and the practical aspects of scale and economics led us to the outcome indicated. Hospitals have their own particular energy demand characteristics and the new generation of low energy ones will be very different from those of more recent times. Taking an approach which is concerned with the energy strategy for the whole hospital is paramount as the overall study has shown. In evaluating solar energy water heating against this basic philosophy one is immediately faced with the fundamental issue which is that large quantities of surplus heat are available during the summer periods from environmental and process sources, within the hospital. Systems have to be installed to recover and use this heat in the winter and midseason and, therefore, the system hardware is available year round and can provide the reduced amounts required during the summer. Any solar system is, therefore, competing for much of its effective operating cycle against recovered heat performance criteria rather than input from a fossil fuel supplementary heat source. These aspects obviously put the solar system into a back seat context when energy contribution and economics are considered. During certain operating periods the recovered heat is obtained at virtually no energy input penalty to the heat pump such as during the summer when the equipment is operating on its refrigeration cycle. The point I am wanting to stress is that we have identified a way forward for designing low energy hospitals, but this is only a small part of the total building stock and the emphasis and impact of solar to other building types needs to be similarly evaluated. Only then will the solar equipment manufacturers be able to discern the requirements of the construction industry and with it the place and purpose that solar energy can play.

6

Ambient Energy in Commercial Buildings

J. Campbell

INTRODUCTION

The commercial and industrial building as we know it today is a relatively new development dating back to the industrial revolution. Before this date most indoor industry was cottage based and office accommodation used rooms that were designed on a domestic scale. With the advent of industrial enterprises that could only be carried out on a large scale it became necessary to mount parallel commercial operations to cope with obtaining the necessary raw materials and organizing the disposition of the finished goods.

The commercial building followed a much more gradual rate of development than the industrial building, the main considerations being internal communications between staff and the owning and operating cost of the building itself.

For many years the energy consumption of a building became a secondary consideration because the largest single sum involved by far in a commercial operation was the wage bill. Although this situation has not changed dramatically in the past few years in real terms, there seem to be two factors which have changed mental attitudes. The cost of fuel is no longer as predictable as it has been historically and in addition, the realization by the public at large that resources are finite has now made waste a moral issue.

When a building is being designed, the engineer responsible makes a value judgement based sometimes on past experience and sometimes on calculation. This assessment is aimed at giving the best value for money to the client, taking into account both capital cost and running cost. Many designers who produce quite good work do not even realize that this is what they are doing. They are using their experience of what was right on a previous job to determine what is right now and deviating slightly on to a new course as the commercial pressures force them. This is a perfectly valid approach in a time of steady change but needs drastic modification, in times of rapidly changing economic circumstances.

For this reason we must start re-examining the design of our buildings. The end product we are attempting to achieve has not changed, it is just that the parameters within which we are attempting to work have been altered, and a change in these will inevitably mean a modification in the end product. For many years we existed in an economic climate which comprised low energy prices and high interest rates.

In a time of low energy cost and high interest rates the initial cost of the building itself became of paramount importance and, in city centres where land costs were high, maximum site utilization was essential. These factors together resulted in the creation of deep plan offices with high lighting levels, and resulting high energy costs.

The approach that must be used when energy costs rise is one of minimizing the energy consumption of a building and ensuring that maximum utilization is made of any fuel consumed. Even more important, we should now start to investigate the feasibility of using other energy sources to supply our buildings. In a lot of cases the basic research has been carried out and the only reason that there has not been any practical application of the results is that the capital cost is high. There is a break-even point at which the energy cost starts to outweigh the capital cost and it is this point that we have to determine.

The term 'use' when applied to ambient energy in conjunction with commercial buildings can be somewhat of a misnomer at the moment. Most commercial and industrial buildings designed today consider ambient energy, but in most cases control and regulation are the first parameters to be considered. The main reason is that the modern commercial building has very high internal heat gains which make it unnecessary to provide heat, but rather make it necessary to limit the extent to which the external environment can affect the internal conditions.

It is a normal practice to break down the utilization of ambient energy into active systems and passive systems. Whereas, however, the active approach always involves the provision of energy, passive systems can again be subdivided into those which result in the provision of heat as a direct contribution to the energy requirements of the building, and those in which skilful design is used to minimize the conventional energy requirements.

To understand the way in which the ambient resources can best be utilized, it is necessary to assess the ways in which energy is used in a building and the methods by which this energy can be produced.

All buildings in a temperate climate, such as that experienced in the United Kingdom require heat, light and power. The proportions in which these are required will vary considerably depending on the use of the building, but in all cases these requirements will have to be met.

Although energy consumption is now at last given a reasonable level of priority in the design stages of a building, it must be borne in mind that salaries are still the major cost in running a building. Energy must not therefore be saved by providing conditions resulting in a drop in efficiency of the building occupants.

ALTERNATIVE SOURCES OF ENERGY

All energy resources are expendable but some more readily so than others, namely the hydro-carbon fuels, coal and oil. We have other energy sources available and these, which are quite numerous, include the following: rivers (natural and artificial), wind, sun, tides, sewage, incineration of rubbish, and nuclear fission.

If we look at the three basic requirements – heat, light and power – we find the following:

(a) Heat can be provided:
   (i) directly by the sun
   (ii) by the combustion of hydrocarbon fuels
   (iii) by the heating of an electrical resistance, as in the case of an electric fire
   (iv) by a heat pump, probably electrically driven.
(b) Artificial light can be provided:
   (i) by electricity
   (ii) by hydrocarbon fuel
(c) Power can be provided:
   (i) electrically
   (ii) mechanically by a rotating shaft.

Examination of (a), (b) and (c) shows that if gas, oil and coal are eliminated from the list, either electrical or mechanical power are required for most operations. Even more enlightening is that the most effective way of producing electricity at the moment is mechanically by a rotating shaft.

In fact it is possible to reduce the problem to one of examining three items:

(a) First order uses of solar radiation – solar collectors, etc.
(b) Second order uses of solar radiation – artificial production of hydrocarbons using photochemistry.
(c) Methods of making a shaft rotate.

From our point of view, the second of these can be ruled out entirely, other than for pure interest. It is a stage removed from our buildings and has, as its aim, the production of a replacement for conventional hydrocarbon fuels.

The use of the sun as a primary heat source is fairly well documented, and it is fairly simple to provide low-grade energy by this method. Using the absorption refrigeration system it is also possible to provide cooling by this method as well. Although the coefficient of performance of absorption units is very low and it is normally difficult to justify their use on anything other than a waste-heat recovery basis, the costs are much easier to justify when the heat is provided by solar energy.

Heating by solar energy does create a problem in that the energy availability is at its lowest when the maximum demand is being made. This means that collectors have to be sized for the worst conditions that will be experienced in winter. For simplicity and cheapness, fixed-plate, matt-black collectors are the most popular type.

Space heating from flat-plate solar collectors via a storage tank or stone storage is also commercially viable in areas similar to Florida, but is obviously dependent on the availability of sunshine, costs of fossil fuels and capital equipment and life of the installation. The University of Florida have an inhabited test house for use as a test ground for commercial equipment and their own experiments.

This now brings us to the third item on the list. How do we make a shaft rotate. This problem has only recently been answered for us by the availability of cheap energy. Before this, rivers and wind were the principal sources of energy. Both of these were used to operate mills and in later years water wheels were used to drive early machinery by a mixture of shaft and belt drives.

Tides have been used by the French for electrical power generation but unfortunately the peak availability of power is governed by the moon and only coincides with maximum demand on rare occasions.

The storage of electrical energy has long been a problem. Batteries are expensive relative to the amount of energy they can store, and have a relatively short life when compared with generating equipment.

The Rankine cycle has come to the forefront as the most efficient way of generating electricity but economics of scale have resulted in turbine sizes steadily increasing over the years. There has, however, been a move in the solar field towards the use of Rankine cycle equipment for the production of cooling. The lower efficiency of these relatively small turbines is not terribly important in running cost terms as the energy source is free but it does mean that very large collector arrays are necessary which result in a high capital cost. Interest rates therefore play a major part in an economic assessment of this type of system.

CURRENT DEVELOPMENTS

Whilst the commercial building has very little requirement for heat because of its internal heat gains it does have a requirement for lighting and it follows that if this can be provided naturally then the consumption of conventional energy can be significantly reduced.

This has led to increased interest in the field of window design. If the psychological arguments for outside awareness are ignored and the window is examined in energy terms, a balance has to be obtained between the energy-saving benefits of daylight and the cost of the heat lost through it. In addition the heat gains in summer can be significantly reduced by the careful use of overhangs. Careful, though, because the overhang reduces the shape factor of the sky from the room and hence the amount of daylight received.

It is not possible to produce a universal solution to the problem of window design. The solution to a particular case is dependent on too many factors which are particular to the building in question. The same factors, however, have to be taken into account in each case and the following examples illustrate some solutions to this problem.

Figure 6.1 shows the Central Electricity Generating Boards Offices at Bedminster Down near Bristol. The perimeter is double glazed with blinds

Figure 6.1. CEGB, Bristol (photograph by Crispin Boyle)

between the panes. High-level glazing is double and unshaded (see Fig. 6.2). This design is the result of optimization studies carried out against the diurnal temperature cycles both for the period May to September and the winter design state, which in turn have shown that the proposal satisfies the required design conditions. The insulation standard for roofs, non-glazed walls, etc. is to achieve 0.6 W/m² °C.

Waste heat is recovered whenever practicable. Vitiated air discharged from the building is passed through a heat exchanger to warm the necessary outside air which is being drawn into the building, whilst other forms of discharged heat are used to heat water in a general heat reclaim circuit.

Task lighting is used as much as possible. The need to control lighting for work stations is inherent in the concept of task lighting.

In winter, when the heating requirements are greatest, use is made of the heat produced by artificial lighting. In summer, when the outside air temperature at

Figure 6.2. CEGB, Bristol. Internal view (photograph by Henk Snoek)

night is low enough, this is used to cool the building. Passing this cool air through a thermal storage system gives the required phase shift to reduce and, in large parts of the building, eliminate the need for mechanical cooling. The thermal storage system installed in the building uses the floor slab to provide the thermal inertia required. Air is passed through hollow concrete floor planks on summer nights to cool the structure to a temperature below the desired room temperature of 22 to 23°C. With the floor slab cooled to 17°C, for example, the daytime air passed through the floor slab will be cooled to about 19°C from the maximum condition of 23°C as the air leaves the air handling plant. This air then absorbs the internally produced heat from lights, people, etc., maintaining a nominally constant room temperature. There are occasions, towards the end of the afternoons of the third and subsequent days of a heatwave, when the temperature rises above the control point. However, the occasions where several successive days reach or exceed the design condition are rare (approximately 73 hours per year).

The capacity for cooling provided by the structure/air heat exchange is necessarily limited. While adequate for office functions where energy inputs to the area are fairly small, it is not adequate for spaces such as computer areas or where constant temperature rooms are required in the laboratory areas. The cooling capacity of the air-floor system is therefore supplemented in these spaces by mechanical cooling.

The refrigeration plant used to provide the chilled water is piped up as a heat pump. This is then used to provide input to the perimeter heating.

For much of the summer there is still surplus heat in the heat reclaim circuit, as is also the case throughout the year during night and weekend hours, when only the computer and telecommunications areas are in operation. A swimming pool is provided for the staff as part of the complex and under these conditions, the surplus heat is supplied to the swimming pool. In this way it is possible to prevent the swimming pool boiler from starting while there is any form of usable heat in the building.

Other subsidiary benefits of this arrangement are that the swimming pool will supply the heat called for by the frost protection system, so that no boiler need start, and as a large thermal store it assists in maintaining a stable condenser water return temperature: an essential requirement of any heat-pump system.

The concept of careful window design is not new. The Victorians were using the north light concept extensively and this provides a very good level of illumination whilst minimizing the effects of direct solar gain and is difficult to improve on. Figure 6.3 shows the application of the north light concept in an office environment. This is a refurbished area in the Cummins diesel-engine factory at Shotts and the lights are now double glazed and heating provided beneath them to reduce the effects of down draught.

Figure 6.4 is the riverside elevation of the new Lloyds building overlooking the Medway at Chatham. The solution used here on the external glazing is very similar to that used on the CEGB building shown referred to previously. This building was also designed to minimize the need for air conditioning and in fact air conditioning is only used in the computer areas. The air conditioning system for the computer area is interesting in that it uses ambient energy to minimize the cooling load. This is achieved by using a floor-level supply system which means that convection gains which have risen to ceiling level are not entrained and circulated around the room but are extracted at high level. The net result is that in maintaining a temperature at breathing level of 22°C, instead of a supply temperature of 14°C and an extract temperature of 22°C being used, temperatures of 18°C and 26°C respectively are quite feasible. In a temperate climate such as that experienced in the UK this reduces the number of hours when some mechanical cooling will be necessary from 2224 hours to 619 hours in a typical year (see Table 6.1). If humidity control is also incorporated the reduction is not quite as dramatic but is still considerable, being from 1673 hours per annum to 894 hours per annum (see Table 6.2).

A building on which this technique has been used exclusively is the Klocknerhaus Building at Duisberg in West Germany. The air-conditioning designers, Schmitt Reuter, have also designed an air supply terminal which can be desk mounted, giving the user some control of the amount of air movement which takes place in his immediate vicinity.

Figure 6.3. Cummins Office

Figure 6.4. Lloyds, Chatham (photograph by Martin Charles)

QUO VADIS

The future is somewhat difficult to forecast as it depends to a large extent on which of the various fuel scenarios being forecast ultimately becomes fact.

These vary radically from there being no shortage at all and it is simply that it will become more difficult to obtain access to the reserves, to one in which the future energy sources will become totally dependent on sun and wind, with the buildings designed to be completely autonomous.

Somewhere in the middle of this range lie the all electric solution, the solar solution and a combination of the two. The latter seems to be a more likely alternative than either of the extremist points of view. Electricity can be generated without the burning of fossil fuel by nuclear fission, and ultimately it is hoped by nuclear fusion, neither of which contribute to the steadily increasing level of CO_2 in the atmosphere.

Two buildings which use this combined approach are the Shenandoah Solar Recreation Center located about 50 km south of Atlanta in Georgia, USA (Fig. 6.5) and the Honeywell Office building in Minneapolis, Minnesota (Fig. 6.6).

At the time of its design and construction, the solar system in the Shenandoah Solar Recreation Center was the largest solar-fired heating and cooling system in the world. The system has a net collector area of 1041 m^2 fitted at a slope of 45°

Table 6.1 CIBS example year for Kew: Hourly distribution of dry bulb temperature

Hours temperature is above the limit value

Limit (°C)	Oct A	Oct B	Nov A	Nov B	Dec A	Dec B	Jan A	Jan B	Feb A	Feb B	March A	March B	April A	April B	May A	May B	June A	June B	July A	July B	Aug A	Aug B	Sept A	Sept B	Year A	Year B
−7	341	744	330	720	341	744	341	744	308	672	341	744	330	720	341	744	330	720	341	744	341	744	330	720	4015	8760
−6	341	744	330	720	341	743	341	744	308	672	341	744	330	720	341	744	330	720	341	744	341	744	330	720	4015	8759
−5	341	744	330	720	341	741	341	744	308	672	339	743	330	720	341	744	330	720	341	744	341	744	330	720	4015	8756
−4	341	744	330	720	339	735	339	741	308	671	337	740	330	720	341	744	330	720	341	744	341	744	330	720	4011	8747
−3	341	744	330	720	336	726	338	739	306	666	337	733	330	720	341	744	330	720	341	744	341	744	330	720	4004	8727
−2	341	744	327	714	335	711	338	729	305	662	323	724	330	720	341	744	330	720	341	744	341	744	330	720	3993	8690
−1	341	744	325	711	328	694	333	711	303	654	323	694	330	720	341	744	330	720	341	744	341	744	330	720	3973	8626
0	341	744	324	705	314	660	307	635	293	620	308	658	330	720	341	744	330	720	341	744	341	744	330	720	3936	8524
1	339	739	322	697	295	620	263	526	265	521	299	619	330	719	341	744	330	720	341	744	341	744	330	720	3868	8322
2	337	733	320	691	265	556	214	420	234	447	286	579	329	713	341	744	330	720	341	744	341	744	330	720	3749	7997
3	335	719	311	668	234	492	181	361	199	362	270	547	321	700	341	744	330	720	341	744	341	744	330	720	3611	7678
4	333	700	298	634	193	395	159	307	147	261	255	523	304	644	341	744	330	720	341	744	341	744	330	720	3471	7344
5	328	671	283	600	139	290	120	233	73	132	243	484	286	577	341	744	330	720	341	744	341	744	330	720	3303	6921
6	315	633	274	576	100	223	91	173	26	46	222	425	256	512	340	737	330	719	341	744	341	744	330	719	3091	6450
7	298	568	260	526	96	203	71	139	11	20	199	371	226	431	338	728	325	715	341	744	341	744	325	715	2938	6026
8	263	481	238	450	88	178	43	95	5	5	174	304	181	344	331	717	318	702	340	741	341	744	318	702	2776	5584
9	217	367	206	358	77	154	22	36	2	2	143	215	149	271	309	697	308	679	337	733	341	744	308	679	2578	5005
10	183	278	166	278	60	123	10	13	0	0	107	148	97	198	278	653	300	640	327	726	337	684	300	640	2347	4466
11	151	219	121	197	45	95	0	0	0	0	83	106	59	124	257	579	277	603	318	709	328	635	277	603	2137	3966
12	118	166	70	124	21	45	0	0	0	0	59	72	34	71	219	497	254	540	296	665	314	564	246	559	1879	3350
13	84	114	27	48	8	16	0	0	0	0	35	43	17	38	175	404	215	474	259	604	286	474	195	460	1616	2747
14	61	77	11	17	0	1	0	0	0	0	26	32	10	19	134	313	174	395	212	520	245	370	154	363	1389	2224
15	43	49	0	0	0	0	0	0	0	0	17	21	5	10	96	240	128	300	161	420	203	273	108	261	1164	1754
16	34	34	0	0	0	0	0	0	0	0	12	16	3	5	63	184	82	232	106	306	165	214	68	192	932	1288
17	20	20	0	0	0	0	0	0	0	0	11	13	0	3	45	126	52	166	59	205	128	159	30	126	705	901
18	8	8	0	0	0	0	0	0	0	0	9	13	0	0	38	84	30	107	35	125	90	109	10	69	496	619
19	2	2	0	0	0	0	0	0	0	0	7	11	0	0	29	60	17	67	19	69	64	76	0	30	328	408
20	0	0	0	0	0	0	0	0	0	0	3	7	0	0	22	50	5	38	9	41	35	41	0	10	213	263
21	0	0	0	0	0	0	0	0	0	0	0	3	0	0	13	39	0	22	2	23	11	12	0	0	138	168
22	0	0	0	0	0	0	0	0	0	0	0	0	0	0	9	28	0	5	0	3	2	2	0	0	74	90
23	0	0	0	0	0	0	0	0	0	0	0	0	0	0	5	16	0	0	0	0	1	1	0	0	30	34
24	0	0	0	0	0	0	0	0	0	0	0	0	0	0	3	14	0	0	0	0	0	0	0	0	11	12
25	0	0	0	0	0	0	0	0	0	0	0	0	0	0	0	6	0	0	0	0	0	0	0	0	6	7
26	0	0	0	0	0	0	0	0	0	0	0	0	0	0	0	3	0	0	0	0	0	0	0	0	3	3
27	0	0	0	0	0	0	0	0	0	0	0	0	0	0	0	0	0	0	0	0	0	0	0	0	0	0

A: daytime hours 0800 to 1800 inclusive.
B: all day hours 0100 to 2400 inclusive.

Table 6.2 CIBS example year for Kew: Hourly distribution of enthalpy

Hours enthalpy is above the limit value

Limit (kJ/kg)	Oct		Nov		Dec		Jan		Feb		March		April		May		June		July		Aug		Sept		Year	
	A	B	A	B	A	B	A	B	A	B	A	B	A	B	A	B	A	B	A	B	A	B	A	B	A	B
−2	341	744	330	720	341	744	341	744	308	672	341	744	330	720	341	744	330	720	341	744	341	744	330	720	4015	8760
0	341	744	330	720	341	742	341	744	308	672	341	743	330	720	341	744	330	720	341	744	341	744	330	720	4015	8757
2	341	744	330	720	338	733	341	744	308	672	337	738	330	720	341	744	330	720	341	744	341	744	330	720	4008	8743
4	341	744	330	720	333	713	338	740	308	670	333	725	330	720	341	744	330	720	341	744	341	744	330	720	3996	8704
6	341	744	327	714	318	667	338	734	305	664	324	693	330	720	341	744	330	720	341	744	341	744	330	720	3966	8608
8	341	744	324	709	292	626	334	711	293	627	297	629	330	720	341	744	330	720	341	744	341	744	330	720	3894	8438
10	338	734	323	697	278	594	278	562	243	513	282	584	330	720	341	744	330	720	341	744	341	744	330	720	3755	8076
12	336	711	320	690	242	495	211	430	207	405	257	541	330	719	341	744	330	720	341	744	341	744	330	720	3586	7680
14	332	675	309	666	193	407	173	360	141	279	247	514	330	716	341	744	330	720	341	744	341	744	330	720	3408	7325
16	323	633	289	619	133	292	128	267	74	178	238	471	318	657	341	740	330	720	341	744	341	744	330	720	3186	6827
18	308	576	269	579	98	219	92	199	38	80	224	419	283	546	341	727	330	720	341	744	341	744	330	720	2995	6330
20	291	491	250	516	88	193	81	160	14	22	185	344	244	452	336	707	330	720	341	744	341	744	330	720	2831	5893
22	249	374	213	421	83	172	68	139	3	6	140	255	193	340	323	656	330	720	341	744	341	744	327	715	2611	5379
24	197	244	164	315	77	152	48	99	0	0	85	162	147	241	292	592	330	708	340	735	341	744	322	691	2343	4785
26	143	174	135	252	61	130	25	45	0	0	64	100	92	139	270	524	327	689	333	715	341	744	310	663	2101	4204
28	103	122	107	194	37	96	12	16	0	0	40	59	54	73	229	434	322	663	318	681	341	741	284	625	1847	3677
30	71	92	77	143	23	46	0	1	0	0	15	16	30	34	175	324	310	615	287	608	336	726	264	561	1588	3096
32	59	58	38	94	17	34	0	0	0	0	6	7	12	13	133	215	283	526	261	532	318	701	213	486	1340	2511
34	43	30	17	34	5	7	0	0	0	0	4	4	2	2	90	137	229	407	225	448	280	626	168	372	1063	1918
36	19	21	10	12	0	0	0	0	0	0	1	1	0	0	52	77	182	312	184	365	235	557	109	166	792	1423
38	13	10	0	0	0	0	0	0	0	0	0	0	0	0	30	42	135	229	162	287	201	460	75	107	616	1056
40	9	5	0	0	0	0	0	0	0	0	0	0	0	0	23	31	97	153	122	205	160	370	50	68	461	742
42	5	1	0	0	0	0	0	0	0	0	0	0	0	0	13	18	66	91	83	131	123	275	30	37	320	479
44	1	1	0	0	0	0	0	0	0	0	0	0	0	0	7	11	38	46	59	84	83	197	12	13	200	264
46	1	0	0	0	0	0	0	0	0	0	0	0	0	0	6	7	15	16	32	36	55	109	0	0	109	126
48	0	0	0	0	0	0	0	0	0	0	0	0	0	0	4	4	0	0	11	13	29	66	0	0	44	49
50	0	0	0	0	0	0	0	0	0	0	0	0	0	0	2	2	0	0	0	2	4	32	0	0	7	8
52	0	0	0	0	0	0	0	0	0	0	0	0	0	0	0	0	0	0	0	0	1	4	0	0	1	1
54	0	0	0	0	0	0	0	0	0	0	0	0	0	0	0	0	0	0	0	0	0	1	0	0	0	0

A: daytime hours 0800 to 1800 inclusive.
B: all day hours 0100 to 2400 inclusive.

Figure 6.5. Shenendoah Solar Recreation Center

facing south augmented by aluminium reflectors facing north at a 36° slope. As can be seen from Fig. 6.5, the building has been very carefully designed to minimize its energy requirements and with a floor space of 5481 m² has a maximum heating load of 121.5 kW and a maximum cooling requirement of only 89 tons. The system is owned and operated by the Shenandoah Development Corporation. The problem of supporting the collectors was solved by using a saw tooth roof structure based upon a long-span open-web truss system. Collector-reflector angles were chosen to optimize heat production in the summertime when collector temperatures sometimes reach 100°C. The entire system is pressurized, and the temperature of water from the collectors has occasionally reached 116°C. Heat storage for the winter is provided by one 60 m³ tank, and chilled water storage is provided by two 113.55 m³ tanks. The HVAC system uses air handling units located at east and west ends of the building. Domestic hot water is provided by a heat exchanger in the collector loop. The system provides 95% of the heating 60% of the cooling and 50% of the domestic hot water requirements.

The Honeywell building has a completely different type of system and the building itself looks more conventional. Instead of an absorption refrigeration system this building incorporates a centrifugal chiller by a Rankine cycle with a turbine circuit using R113. The collectors shown in Fig. 6.6 are located on the top floor of the car park and have a total area of 1890 m². They are of the single axis type with an off axis parabolic reflector with a concentration ratio of 40:1. The cooling system comprises two 100-ton chillers each requiring a shaft input of 85 HP to achieve this output. With the cost of concentrating collectors being

Figure 6.6. Solar collectors, Honeywell, Minneapolis

extremely high it is important to have as high a coefficient of performance as possible and in this case 5.46 has been achieved. System performance so far indicates that when experimentation has been completed the system will provide 100% of the domestic hot water 80% of the cooling and 50% of the heating loads.

An area of solar collection which shows great promise in practical terms is that of photo-voltaics. Figure 6.7 shows a photo-voltaic array currently being evaluated by the Georgia Power Company. The cost of this type of cell is now in the region of $6 per peak watt for large orders. The goal of the US Department of Energy is to reduce this to less than 70 cents per peak watt by 1986. It is probable that this date will be improved on. Because of the high cost of the silicone used in the cells currently available, it is possible that the breakthrough, when it comes, will be with a different material.

Although there will be definite advantages to be obtained in a temperate climate like ours from the use of photo-voltaics, the benefits in a warmer climate will be twofold. The peak electrical load in warm climates normally occurs in summer co-inciding with the peak air-conditioning load. Photo-voltaics therefore not only reduce the amount of fossil fuel consumed but also improve the load factor on the conventional power plant.

ECONOMIC CONSIDERATIONS

Methods of assessing the economics of a given arrangement for determining the payback period for a given investment are legion but there are two approaches

Figure 6.7. Photo-voltaic array, Georgia, USA

Figure 6.8. Solar collectors providing domestic hot water at Pedigree Pet Foods

that are simple to use and give a good indication of the comparative overall economic viability of a range of solutions.

The first of these methods consists of obtaining the annual cost of owning and operating the different arrangements. This consists of adding the annual cost of amortizing the principal sum of the running.

The annual cost can be calculated using the formula:

$$P = \frac{R}{1 - \{1/[1+(R/100)]^n\}}$$

where P is the percentage proportion of the capital sum that must be paid annually for the term of the loan, R is the rate of interest and n is the term of the loan. The annual running cost for each solution is then added to the annual capital cost and the solution which gives the lowest answer is the most economic.

The other solution which is of interest is the accumulated cost method. In this case the accumulated cost (capital and running cost) is plotted graphically against time. If two solutions are plotted on the same graph it is possible to see how long it takes to recover the additional capital outlay. An example is shown in Fig. 6.9.

CONCLUSIONS

The economic factors governing the choice of an ambient energy system are important and will continue to be so. There are signs, however, that attitudes are

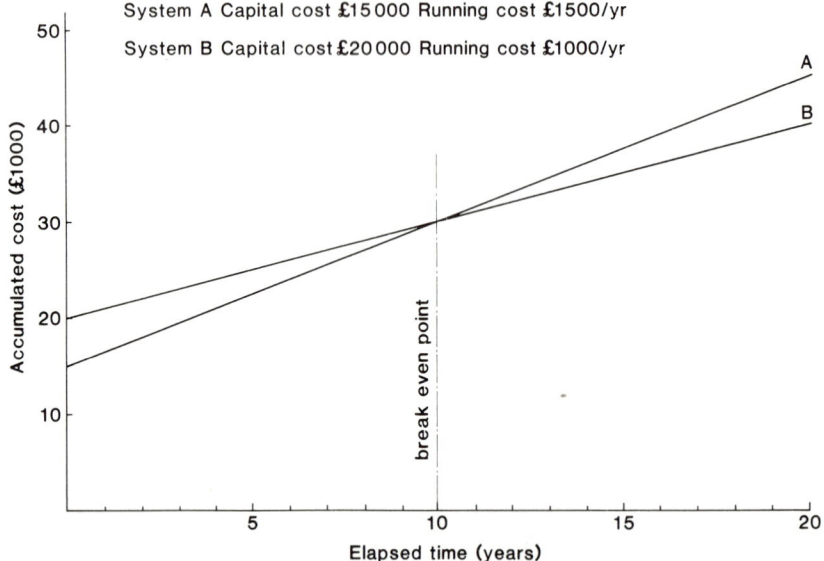

Figure 6.9. Economic considerations

already changing slightly in the USA and that payback periods do not have to be in the three-year area. In fact the photo-voltaic industry there has built up to a multi-million dollar annual turnover without any possibility of amortization at current interest rates.

In the UK a solar-heated, domestic hot-water installation designed to provide heat in the summer months – when a gas boiler may be operating at a poor part-load efficiency – can be shown to have a payback period of between seven and ten years depending on the fuel-cost inflation figure chosen. As the average period of house occupancy is in the region of seven years, only people who know they will be staying in one house for more than this period would consider it worth investing initially. This situation is starting to change a little now. Solar HWS systems are becoming desirable, possibly even a status symbol, and as a result capital invested can be recovered from the sale of the property.

The situation with the commercial or industrial client is somewhat different in that a profitable organization capital expenditure can be discounted against tax, shortening the payback period.

Solar air conditioning is still in the demonstration project stage. It is possible to produce solar systems that work satisfactorily but an all solar system is not likely to pay for itself in the life of the installation at the moment.

The passive techniques of window shading and free cooling with air at the outside ambient temperature are nearly always valid economically and should be used even when the remainder of the energy is to be provided by an active system.

An area always worth consideration at the moment is that of the heat pump or double bundle condenser. Many installations are now being designed with thermal storage vessels into which waste heat is dumped so that it can be recovered the following morning if heating is required.

Possibly the best prospects for active input lie with photo-voltaic conversion at the moment but only time will tell.

DISCUSSION OF CHAPTER 6

Dr J. C. McVeigh (Brighton Polytechnic). Mr Campbell suggested that solar air conditioning was only at the demonstration stage. I would give a slightly different emphasis: *While* some solar air conditioning systems are in the demonstration project state, *nevertheless*, it is possible to provide solar systems that are satisfactory. Solar air conditioning systems can be bought off the shelf after all – a lithium-bromide system, that can be linked to existing off-the-shelf tracking solar systems or evacuated tube ones. I would agree that 'an all Solar system is not likely to pay for itself' although some people are now showing that systems are competitive now with conventional ones, depending on the predicted future pricing strategy for electricity. Solar air-conditioning systems are being preferred by some clients now so it seems likely that there will be an increasing trend towards solar air-conditioning systems in a growing number of applications.

J. Campbell (Ove Arup & Partners). Dr McVeigh's general thesis is accepted but present electrical costs, capital costs of the panel, etc. are such that fuel cost must be high if a solar system is to be made to pay.

W. Carson (University of Glasgow). What method would Mr Campbell use to utilize the energy from photo-voltaic cells for cooling.

J. Campbell. We had considerable debate as to whether the photo-voltaic array should be coupled directly to the refrigeration equipment or whether it is better simply to connect the photo-voltaic array into the normal electrical system, and just draw from the electrical supply in the conventional way. The sun's availability certainly in the hot climates is such that it will do peak lopping to the supply drawn from local power stations. That has, therefore, been seen to be the logical use of the photovoltaic array.

7

Ambient Energy Utilization at the National Centre for Alternative Technology

John Willoughby and Robert W. Todd

INTRODUCTION

This chapter describes an energy audit carried out at the Centre for Alternative Technology between November 1979 and November 1980. The results identify the amount of energy supplied by each of eight sources and how the energy is distributed between end uses. Over this period 38% of end-use energy comes from renewable sources.

The results form the basis of a strategy for future developments which, if implemented, would provide over 60% of the end-use energy requirement from renewable sources.

The Centre for Alternative Technology is a project which aims to demonstrate ideas and methods for achieving a future which is sustainable in the long term [1–3]. This not only involves utilizing, demonstrating and developing various methods of ambient energy production but also investigating food production and human resources.

Initiated in 1972, the Centre occupies a disused slate quarry of 12 ha near Machynlleth on the Powys–Gwynedd border. The site has no mains services. Around twenty-five people work at the Centre with eighteen people permanently resident on the site, occupying about 400 m² of residential accommodation. In addition there is a further 800 m² of non-domestic buildings – exhibition areas, shop, restaurant, a conference centre and workshops with a throughput of 60 000 visitors per year.

The main objectives of the twelve-month audit were:

(a) To assess the amount of energy needed to sustain an energy-conscious community.

(b) To highlight areas where economies and a more rational use of resources could be achieved.

(c) To help define a strategy for future developments of energy technology at the Centre.

Many of the energy-related devices and systems are still at the development stage and some are not used continually. The results of the audit therefore apply to the particular situation at the Centre and are not necessarily representative of the potential of the energy sources or techniques used.

ENERGY SUPPLY AND DEMAND

Energy used in connection with activities at the Centre is shown in Table 7.1. The audit has attempted to identify how much energy was supplied by each source and to what use it was put. In addition it was felt important to ascertain whether the use was domestic or non-domestic.

Table 7.1 Energy sources and uses

Source	Use
Wind power	Electricity for lighting, power and water heating
Water power	Electricity for lighting, power and water heating
Solar energy	Space and water heating
Methane	Water heating (in the dairy)
Wood	Space and water heating
Coal	Space and water heating
LPG	Cooking, water heating, limited space heating and electricity generation
Petrol	Transport (domestic usage excluded from the audit)

The site electricity system [4]

The site electricity system is shown diagrammatically in Fig. 7.1. Pelton-type water turbines supply electricity directly at 240 V a.c. and can be used to charge the 40 kW h battery store. The aero generators feed the system only via battery storage at 24 V or 110 V d.c.

External lighting and lighting in the central buildings are supplied at 24 V d.c. In addition some machinery and control equipment operate at 24 V d.c. A 1 kVA static sinewave invertor is used to produce 240 V a.c. from the battery store for a precision mains supply (used to run recording equipment continuously) and for general use if required.

During long, dry, windless periods a LPG-driven, stand-by generator is used. This provides 2.5 kW at 24 V for battery charging and up to 10 kVA at 240 V for

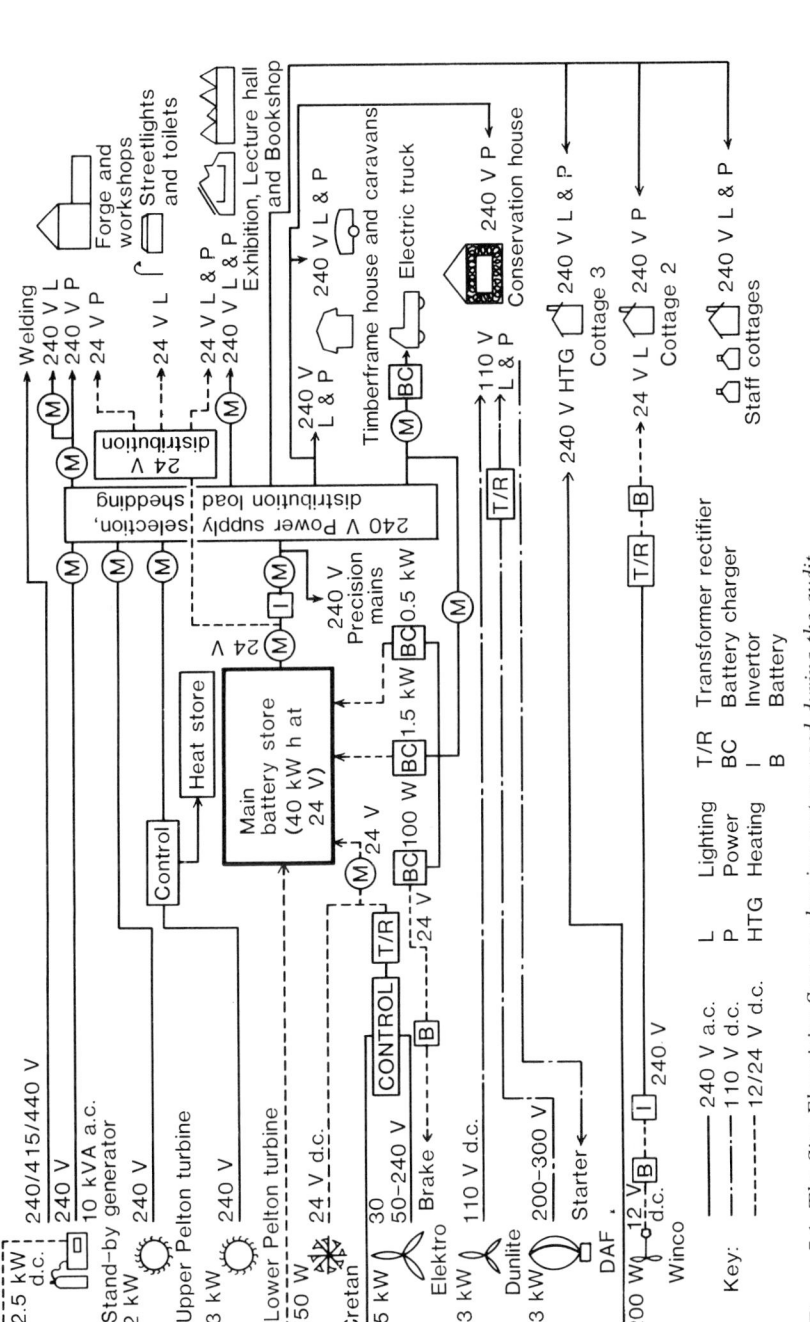

Figure 7.1. The Site Electricity System showing meters used during the audit

general use. Also it can provide power for electric arc welding and to operate a three-phase lathe and milling machine. In times of energy shortage an automatic priority system sheds less important loads to conserve battery energy. The metering arrangements used during the audit are shown in Fig. 7.1. Meter readings were taken every week which allowed an assessment to be made of the energy supplied from the various sources and the final use.

Wind power

Five aerogenerators supply electricity to the site. A 3-kW 'Dunlite' and a 3-kW DAF vertical-axis machine are used in conjunction with the Water Conservation House (Fig. 7.2) providing more than enough energy at 110 V for lighting and power consumption. A small 200 W 'Winco' provides 24 V lighting for one of the staff cottages. The 5 kW 'Elektro' (Fig. 7.3) and the 750 W 'Cretan' are usually

Figure 7.2. The 3 kW Dunlite Aerogenerator poorly sited as an exhibit in front of the Wates Conservation House

Figure 7.3. Three aerogenerators on a site 200 m from the central buildings, from left to right: The DAF 3 kW vertical axis, the 200 W Winco and the 5 kW Elektro

used to charge the main battery store, although the 'Elektro' can be used for heating and hot water in another cottage [6]. These power ratings are based on a rated wind speed of 10 m/s or 11 m/s. Although the Elektro has a rated output of 5 kW at 10 m/s the a.c. to d.c. transformer-rectifier is only rated at 1.5 kW which effectively reduces the rated windspeed when charging batteries (see Appendix 1).

A system for recording the output from the Elektro and/or the Cretan was developed using a purpose-built d.c. amp-hour meter (Fig. 7.4) details of which are given in Appendix 1. The meter was installed in April 1980. It was subsequently rendered inoperative for two short periods due to excess voltage surges (one of which was caused by a lightning strike). However, sufficient data was recorded to allow an estimate of the wind energy supplied to be made by comparing the meter readings with continuous wind speed records from a cup anemometer sited adjacent to the windmills. The method employed is discussed in the Appendix (p. 176).

Water power

A reservoir with a storage capacity of around 4500 m³ feeds a Pelton wheel (Fig. 7.5) with a rated output of 2 kW at a head of 30 m. During the course of audit the system was extended to allow water from this turbine to be fed to another more modern 3 kW Pelton wheel (Fig. 7.6) over another head of 40 m.

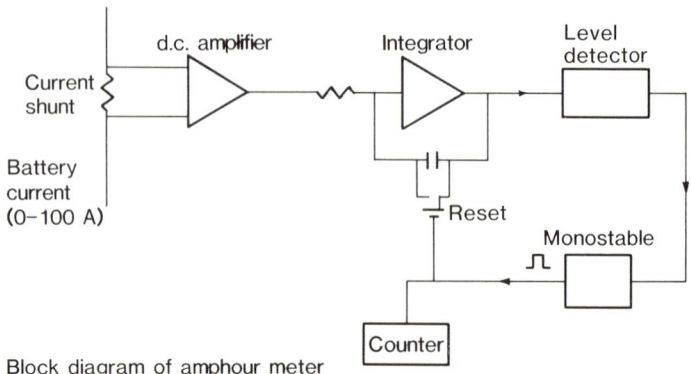

Block diagram of amphour meter

Figure 7.4. Block diagram of the amp-hour meter used to monitor the output from the Elektro and/or the Cretan aerogenerators

Figure 7.5. The upper Pelton turbine. A refurbished 1930s machine with a rated output of 2 kW

Figure 7.6. *The lower Pelton turbine. A new vertical-axis, four-jet machine with a rate output of 3 kW*

The reservoir thus represents a 400 kW h reserve of electricity. The lower Pelton wheel came 'on stream' during June 1980. Both turbines supply electricity at 240 V a.c. The output of each turbine was monitored using standard electricity meters (Fig. 7.1).

Solar energy

Apart from a small array of silicon solar cells which are used to power a pump on a solar hot-water system, solar energy is collected for space and water heating by flat-plate collectors which use water–glycol as the working fluid. The largest system uses a 100 m² trickle-type collector and a 90 m³ interseasonal water store to heat the main 200 m² floor area exhibition building [7] (Fig. 7.7). Five smaller solar systems, with a total collector area of 25 m² supply domestic hot water to

Figure 7.7. The exhibition building is heated with 100 m² of double-glazed, trickle-type collector. 70 m² are on the roof of the building at a tilt of 30°. The remaining collectors have a tilt of 70° to improve the performance during the winter months

Figure 7.8. A 4.4 m² array of solar panels built to a low cost design developed at the Centre. The thermosyphon system preheats domestic hot water for one of the staff cottages

staff cottages and caravans. These use either proprietary panels or those of the type developed at the Centre [8] (Fig. 7.8). With the exception of one installation with 10 m² collector area and a separate preheat tank, the systems all supply hot water during the summer months using one-tank systems which in winter are heated indirectly with solid fuel. The solar energy used in the main exhibition building has been estimated from monitoring carried out over two years [9]. Samples of the results from this survey are illustrated in Fig. 7.9 and show how the system performed over three week-long periods at different times of year.

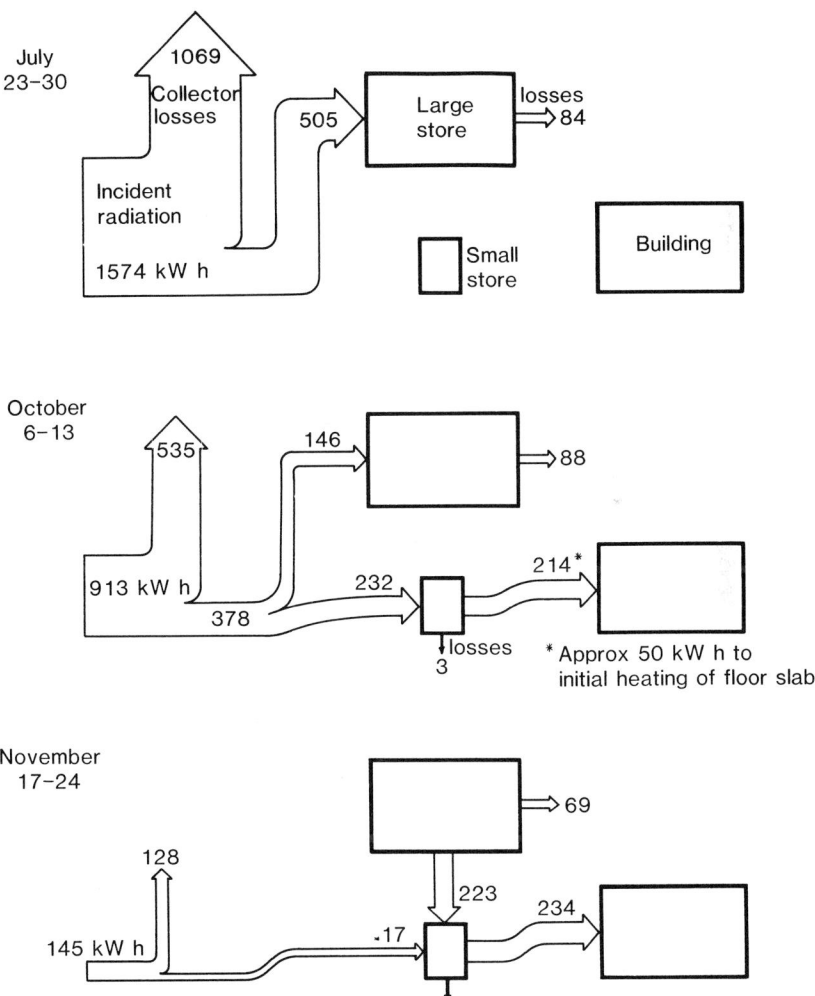

Figure 7.9. The performance of the solar-heated exhibition building for a typical 3-week period

The energy supplied for domestic use is difficult to measure precisely without the use of expensive instrumentation. An estimate was made from measurements of insolation using a Kipp and Zonen solarimeter and integrator. It was estimated that the useful energy output from the domestic solar water heaters would be approximately 20% of the incident radiation [10].

Methane

A small experimental digester is being developed to provide methane for water heating in a small dairy at the Centre. During the audit it was used for twenty weeks. The energy used was estimated as a proportion of the total gas used in the dairy.

Wood

The forestry land around the Centre provides a cheap and plentiful source of fuel. It is used in wood-burning stoves which provide space and hot-water heating in both domestic and non-domestic buildings.

The most accurate method to estimate the energy supplied from this source would have been to weigh the wood as it was being used. However, with eight appliances using wood from three different stores it was felt that this was practically impossible.

Therefore, the stacked volume of timber was measured as it was put into the wood stores. A factor of between 0.4 and 0.6 was applied to this stacked volume to find the net volume. Samples were taken to estimate the density and moisture content. An average moisture content of 30% was found. Using this figure the calorific value was taken to be 3.9 kW h/kg [11].

Coal

Anthracite and a small amount of house coal are used for space and water heating in two cottages, the lecture hall, restaurant and bookshop (Fig. 7.10). Deliveries to the site were recorded and periodic stock checks were made to assess the rate of consumption.

Liquid petroleum gas

Butane is used for domestic cooking and water heating. Propane is used for non-domestic catering and to fuel the stand-by generator (Fig. 7.1). In addition a propane-fired boiler has been installed to provide heating and hot water in the Wates Conservation House. This is not considered to be a very satisfactory solution but was installed because of problems with the heat pump and to provide an easily monitored energy source.

Figure 7.10. Part of the new building which was completed during the course of the audit. The ground floor contains a lecture room, kitchens, restaurant and bookshop with a meeting room and bedrooms situated at first-floor level

The amount of energy used was monitored by noting bottle changes on record cards attached to each of the bottles.

Petroleum

A small petrol-driven van is used in conjunction with an electric truck to transport goods to and from the Centre. The petrol consumption and mileage were recorded and the electricity used to charge the truck batteries was metered to give an estimate of the energy used for transport.

Table 7.2 shows the calorific values of the fuels used during the audit.

RESULTS

Table 7.3 shows the energy delivered to the site by source. The majority of this energy was used for space and water heating, and cooking. Most electricity is used for lighting and electric motors and only surplus electricity which cannot be stored in batteries is used for heating. Figure 7.11 shows the pattern of supply in these three sectors over the year.

Table 7.2 Calorific values

Fuel	Calorific value
Wood	3.9 kW h/kg
Anthracite	9.4 kW h/kg
Housecoal	7.8 kW h/kg
Butane	12.7 kW h/kg
Propane	12.8 kW h/kg
Petrol	8.9 kW h/kg (40.4 kW h/gall)

Table 7.3 Energy delivered to the site (November 1979–October 1980)

Source	kW h (delivered)
Wind	710
Water	7 371
Solar	9 645
Methane	157
Wood	42 296
Coal	90 405
LPG	54 744
Petrol	4 447
Total	209 775

An assessment of the useful energy was made by applying various system efficiencies to the delivered energy figures. Figure 7.2 shows the useful energy by source and by sector.

DISCUSSION

The figures show that energy usage on the site was not excessive. The domestic energy use (65 696 kW h/annum – useful) was approximately half to two-thirds of the average national figure that might be expected for a population of this size [12–14]. Figure 7.12 shows that nearly 15% of the useful energy was supplied by ambient sources and that 38% of the total was from renewable sources. About 80% of the electricity supply was from renewable sources. The figures also highlight three main areas of concern.

(a) For various reasons (detailed below) the full potential of the ambient sources was not being exploited.

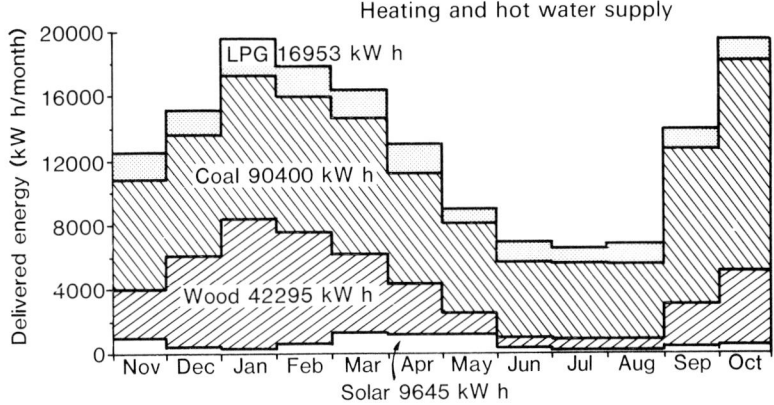

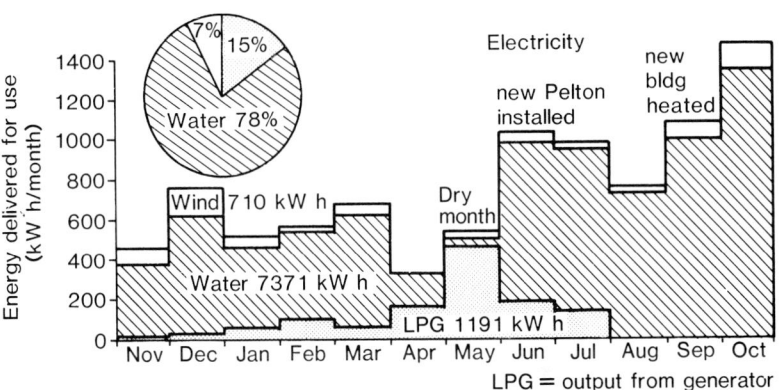

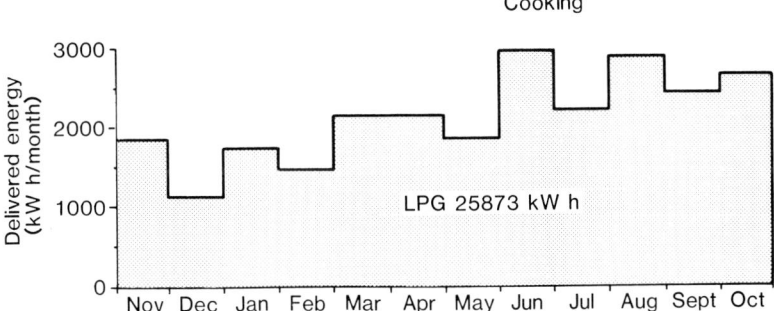

Figure 7.11. The annual distribution of delivered energy

BY SOURCE

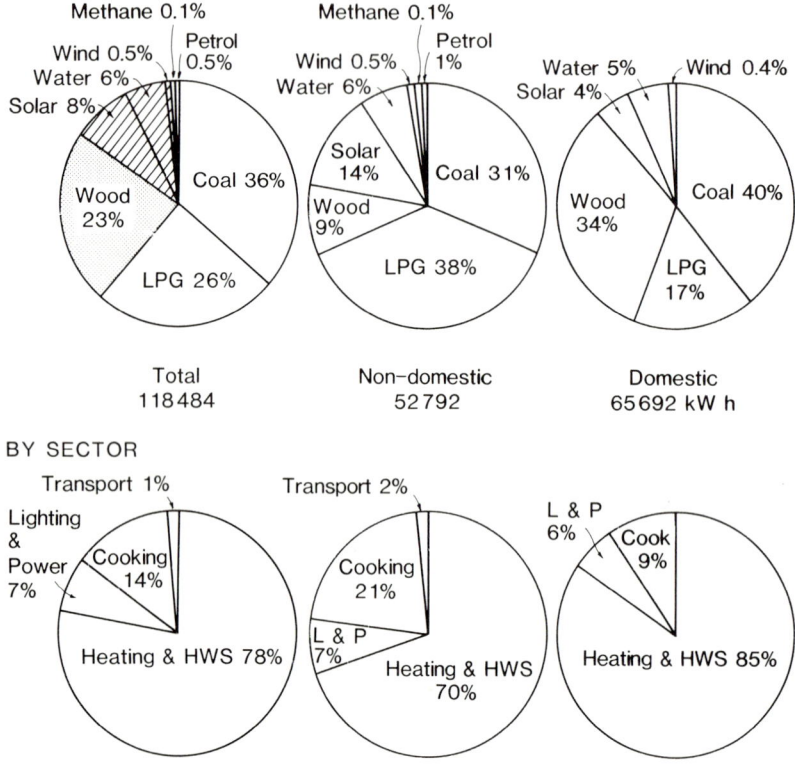

Figure 7.12. *The analysis of useful energy by source and sector*

(b) The anthracite being used for domestic heating was being burnt inefficiently in oversized appliances.
(c) While LPG represents only 26% of the total useful energy the cost of providing the 31 088 kW h/annum (useful) was more than the total cost of all the other fuels purchased.

Each of the energy sources was examined to develop a strategy for the future which would increase the use of ambient energy and reduce the reliance on imported fuels. The proposals were confined to developments which are being undertaken at present or are proposed in the short to medium term.

Wind

The contribution from the aerogenerators was disappointingly low during the audit. This was due to four main factors:

(a) Mechanical and electrical failures including lightning damage.

(b) Poor matching between the Elektro and its load.
(c) Low wind-speed sites, particularly the Dunlite.
(d) Design faults – Elektro blades, DAF [5].

It is considered particularly important that the wind-energy component is increased in order to reduce the reliance on the LPG driven stand-by generator. It is proposed to improve the performance of the Elektro by fitting new blades and improving load matching and control. These modifications should increase the output of this aerogenerator by 1000 kW h/annum (see Appendix, p. 177).

Since a major overhaul during April 1980 the Dunlite has performed as well as can be expected on its present site. Wind speeds recorded over the past year on another hill-top site adjacent to the Centre indicate that by resiting this machine its output would increase to around 2200 kW h/annum. At present the Dunlite is used to provide lighting for the Wates Conservation House which, except in long, low-wind periods, does not fully utilize its output even on the present sheltered site.

It is intended to incorporate the batteries at present in the Wates House into an enlarged central battery store to give a total storage capacity of 80 kW h. The output from the Dunlite would be fed to this store thereby increasing its useful contribution.

The DAF Darrieus turbine has not yet performed satisfactorily over a significant period due to design faults. Although some of the problems have been solved, its net energy output so far has been insignificant.

Plans are in hand to use the new hill-top site for another larger aerogenerator. The details are not yet finalized but machines based on designs of ITDG [15] and Imperial College [16] are under consideration. A relatively low tip speed ratio ($\lambda = 3$ or 4) design is favoured for its good starting characteristics in light winds, for ease of design and construction and high reliability (see Appendix).

It is estimated that these measures would result in a contribution from wind power of at least 18 000 kW h/annum (useful).

Water

At present water power provides 78% of the total useful electricity supply. It is the most reliable and consistent of all the ambient sources. The new lower Pelton wheel is still being commissioned and when the optimum output is achieved it is expected to contribute at least 11 400 kW h (useful) over a full year. The surplus electricity is at present used for heating domestic hot water, but may be fed to a heat store when this is installed. If a new, more efficient generator and better load control facilities were fitted to the upper Pelton its output could be increased from its present 3879 to 6630 kW h/annum (useful).

Solar

Solar energy provides 7300 kW h/annum (useful) for space heating and 2345 kW h/annum for domestic hot-water heating. It is proposed to increase the space-heating contribution by building solar greenhouses on the south side of two of the staff cottages and a solar wall on a third cottage. The greenhouses should yield 3400 kW h/annum. The solar wall will be based on the design developed by Lee and Fitzgerald [17] rather than the more usual Trombe wall and should provide 1600 kW h/annum which is approximately 20% of the space-heating load of the cottage.

There is an ongoing programme of installing solar hot-water systems which will be continued with particular attention being paid to reducing the amount of LPG needed for water heating. Domestically there is a potential to provide at least another 2700 kW h/annum for hot-water heating. A high priority is to use solar energy to augment the hot water used for non-domestic catering. This use is particularly well suited to solar heating because 68% of the load occurs between March and September. It is estimated that a solar system providing pre-heating for non-domestic hot water could produce 1800 kW h of useful energy per year.

Methane

The performance of the methane digestor has recently been improved. Solar panels have been installed to increase the temperature in the digester. During some of the trial periods more than enough methane has been produced to satisfy the modest needs of the dairy. If this improved performance can be maintained, and 80% of the water-heating requirement in the dairy can be met by the methane plant, the contribution from this source will increase to 370 kW h/annum.

Wood

Wood costs the Centre 0.43 p/kW h (delivered) (£17/tonne including the cost of collecting, sawing and stacking). It is at present the cheapest of the bought fuels and in the long term it is hoped to develop coppices on land at the Centre which would provide all the wood needed for fuel. Therefore, it is proposed to increase the use of wood for space and water heating. The most immediate requirement is to install efficient wood-burning stoves in place of the inefficient anthracite burning appliances at present being used to heat one caravan and one of the staff cottages.

The use of passive solar energy and water heating by the Pelton turbine dump load will, however, reduce the need for wood elsewhere. The net result would be an increase of 4020 kW h to a total contribution from wood of 31 512 kW h/annum.

Coal

The useful energy supplied by coal over the period of the audit is shown in Table 7.4.

It is the intention in the future to make a saving of 19 397 kW h/annum (useful)

Table 7.4 Coal usage (November 1979–October 1980)

Type	Use	Useful energy (kW h)
Anthracite (stovesse)	Domestic heating and hot-water supply	19 397
Anthracite grains	Non-domestic heating	16 267
House coal	Domestic heating and hot-water supply	6 647
Total		42 311

by substituting wood for anthracite as described above. The recently completed building, housing the bookshop, restaurant, kitchens and lecture facilities is heated with two Pither stoves burning anthracite grains. The predicted energy consumption is around 39 000 kW h/annum. It is expected that with the increased wind and hydro contributions there will be just under 9000 kW h of spare electricity which will be fed to a space-heating store in this building. The net result would be a requirement for coal of 37 750 kW h/annum (useful).

Liquid petroleum gas

Table 7.5 shows the distribution of useful energy from LPG:

Table 7.5 Liquid petroleum gas use (November 1979–October 1980)

Use	Useful energy (kW h)
Stand-by generator	1 192
Cooking	16 817
Hot water	7 591
Space heating	1 111
Wates House (space and hot-water supply)	3 990
Dairy	387
Total	31 088

Over 50% of the LPG usage is for cooking. It is unlikely that anything can be done to reduce this load in the short to medium term. However, major savings in the other areas are proposed as follows:

(a) The increased wind and water power and battery storage should provide sufficient electricity to reduce the need for the generator except for heavy usage such as welding and lathework requiring a supply of over 3 kVA. It was estimated that this requirement amounted to only 130 kW h during the audit.

(b) Use for water heating will be reduced by solar water heaters and resistance heating by surplus electricity and an increased methane production.

(c) LPG space heating will reduce to negligible proportions as the programme of insulation and the installation of new heating appliances are completed.

(d) It is proposed to substitute an electricaire unit for the LPG fired boiler which at present heats the Wates Conservation House. This load is well matched to the improved performance expected from the Dunlite aerogenerator (see 'Wind' section).

Petrol

It is intended to continue to rely on the electric truck for the majority of non-domestic transport. Petrol-driven transport will have to be used for longer journeys but it is difficult to estimate the extent of this usage.

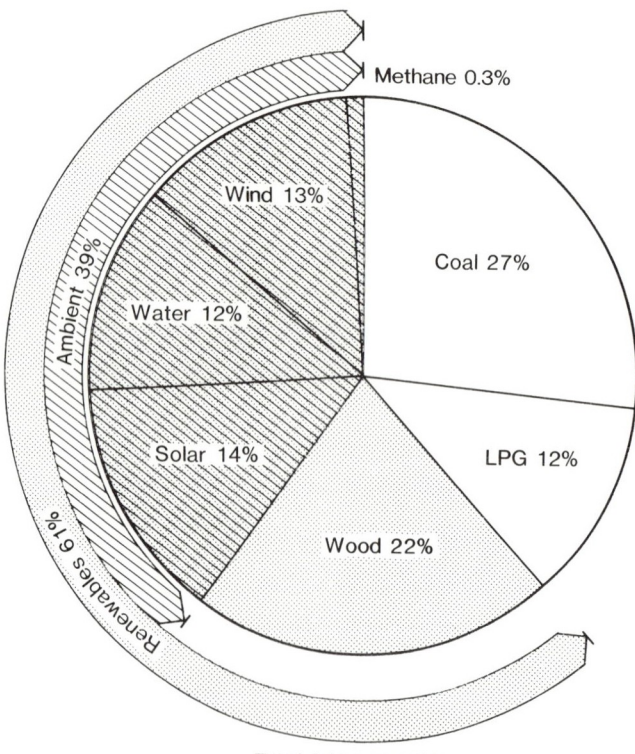

Total 141 112 kW h

Figure 7.13. The estimated pattern of useful energy supply attainable in the short to medium term

CONCLUSION

Figure 7.13 shows an estimate of the useful energy supply which should be attainable in the short to medium term. It has been assumed the proposals set out above have been completed and that the restaurant–bookshop complex is functioning all year round. In addition a 20% increase in electricity consumption has been allowed for.

It can be seen from this figure that it is not unrealistic to expect that the Centre will move to a position where 39% of its energy is supplied from ambient energy and 61% from renewable sources.

REFERENCES

1 Thring, M. W. (1980) *The Engineer's Conscience*, Northgate Publishing, London.
2 Dunn, P. D. (1978) *Appropriate Technology*, Macmillan, London.
3 Lovins, Amory B. (1977) *Soft Energy Paths*, Pelican, London.
4. Todd, R. W. (1980) *Energy and Buildings at the Centre for Alternative Technology*, Technical Information Report No. 1, Centre for Alternative Technology, Machynlleth, Wales.
5. Todd, R. W. (1979) *Experience with Wind Power*, Centre for Alternative Technology, Machynlleth, Wales.
6. Todd, R. W. (1978) Low energy housing – National Centre for Alternative Technology. *Ambient Energy and Building Design* (ed. J. E. Randall), Construction Press, Harlow, pp. 83–96.
7. Todd, R. W. (1978) A Solar Heating System with Interseasonal Storage. *UK ISES Conference C15*. UK Branch, International Solar Energy Society, London.
8. Centre for Alternative Technology (1978) *Solar Water Heater*, DIY Plan No. 4, Centre for Alternative Technology, Machynlleth, Wales.
9. Commission for European Communities (1979) *Performance monitoring of Solar Heating Systems in Dwellings*, Commission of European Communities, EEC.
10. British Standards Institute (1980) *Code of Practice for Solar Heating Systems for Domestic Hot Water*, BS 5918, London.
11. Country College (1977) *Wood Fuel in Britain*, Country College, Alford, Lincolnshire, UK.
12. Barrett, M. *A Dynamic Physical Energy Model of the UK*, PhD Thesis, Open University Energy Research Group, Milton Keynes (in preparation).
13. Leach, G., Lewis, C., Romig, S., Van Buren, A. and Foley, G. (1979) *A Low Energy Strategy for the United Kingdom*, Science Reviews and International Institute for Environmental Development, London.
14. Bush, R. P. and Matthews, B. J. (1980) *The Pattern of Energy Use in the UK*, ETSU R7, Department of Energy, Harwell, UK.
15. Dunn, P. D. and Ersa, T. (1980) *Performance Measurements on 6 m Horizontal Axis Windmill*, British Wind Energy Association Annual Workshop, Cranfield, UK.
16. Freris, L. L., Bolton, H., Buehring, I. K. and Nicodemon, V. C. (1979) A low cost wind energy conversion system. *Future Energy Concepts IEE Conf.* **171**, Institution of Electrical Engineers, London.
17. Lee J. B. G. and Fitzgerald, D. (1980) Comparison of Physical and Mathematical Models of Solar Walls, *UK ISES Conference C24*, UK Branch, International Solar Energy Society, London.

APPENDIX: WIND ENERGY

Three of the wind machines (the Elektro, the DAF Darrieus and the Winco Windcharger) at the Centre are sited quite close together 200 m from the main buildings on high ground.

Monitoring

Wind records have been kept for three years on this site using a cup counter anemometer which also provides an instantaneous velocity signal in the control room. This feeds a single-channel chart recorder and can also be used for control purposes (e.g. starting the Darrieus machine when conditions are suitable) and for monitoring the performance of the wind generators. Wind records have also been kept for periods of up to a year at other nearby sites and these have revealed a much better site some 400 m from the main buildings.

All the operating wind generators feed electricity to battery storage and some difficulty was experienced in finding a cheap but reasonably accurate method of measuring the energy fed to the batteries. A conventional Faraday disc d.c. amp-hour meter was tested but found to have an inadequate dynamic range, i.e. friction effects caused it to ignore small current levels completely. The problem was solved by building a special purpose two channel electronic amp-hour meter (Fig. 7.4). This device has a drift of approximately $\pm 0.1\%$ of full-scale input, so if the average of the current being measured is one-tenth of the full scale, the accuracy of the amp-hour output will be $\pm 1\%$. The minimum input threshold is suitably low, around 0.1% of full-scale input. This device has been used to monitor battery load and the output of the Elektro wind generator.

Results

Results from these measurements on the Elektro (5 kW, 5-metre, 3-phase a.c.) machine are only available for a short period due to various practical problems, not least of which was the far-reaching effects of a lightning strike. Results were recorded intermittently over seventeen weeks. However, data over whole monthly periods was collected only during August, September and October 1980. These measurements are shown in Fig. 7.14. A theoretical simulation using the measured wind data suggested that the annual output from the Elektro would be 530 kW h. This figure would be further reduced by shutdown periods due to faults. Also this prediction is in fair agreement with an annual estimate of 450 kW h extrapolating the average readings for the seventeen weeks over the whole year. The latter estimate made no allowance for the higher output which would be expected during the windier months of November and December 1979.

The measured output of the Elektro has been consistently below expectations: the power produced at the rated wind speed is, according to our measurements,

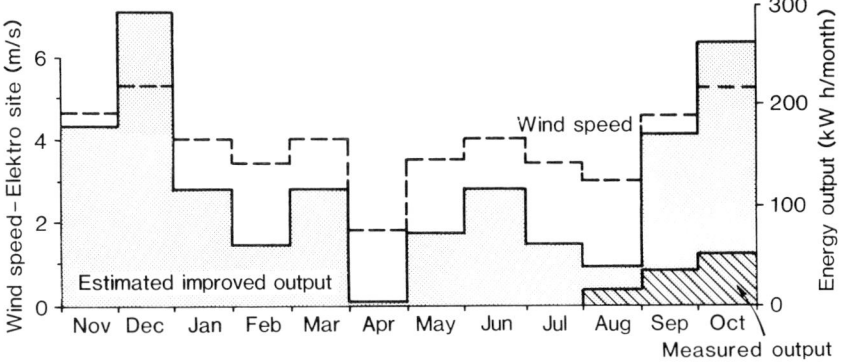

Figure 7.14. The annual wind speed and output from the Elektro aerogenerator

only around half of the published figure of 5 kW, although 5 kW is available at higher wind speeds.

The measurements shown are also adversely affected by the inferior replacement blades in use on the Elektro during the measuring period (these were made necessary by lightning damage). In addition the output is reduced by premature furling in high wind speeds and by poor matching between the generator and battery characteristics. The power drawn from the generator should vary approximately as $(RPM)^3$ but the arrangement at present in use does not maintain this over the necessary speed range. The result is a fall in tip-speed ratio and hence in power coefficient at the higher windspeeds as the rotor fails to reach the appropriate RPM.

Improvements

Work is in hand on an improved matching circuit using a transistor chopper. The estimated improved output shown for the Elektro in Fig. 7.14 assumes both standard blades and improved matching and furling control. These improvements result in an estimated annual output of 1500 kW h.

Only the first 1.5 kW of output from the Elektro is converted to d.c. for battery charging. The infrequent occurrence of powers greater than 1.5 kW (on this site) particularly in summer, makes the additional cost of a larger conversion unit difficult to justify. Excess power above 1.5 kW can be fed to thermal storage. This demonstrates an important principle in multi-use wind systems. Priority should be given to loads for which storage is difficult or expensive, e.g. electricity for lighting etc. and the more intermittent higher power levels should be used for heating or water-pumping loads with their cheaper storage costs, thus minimizing the total system cost.

To increase the wind energy contribution, the main battery store has recently been enlarged to 80 kW h and the installation of a new 10 m diameter wind

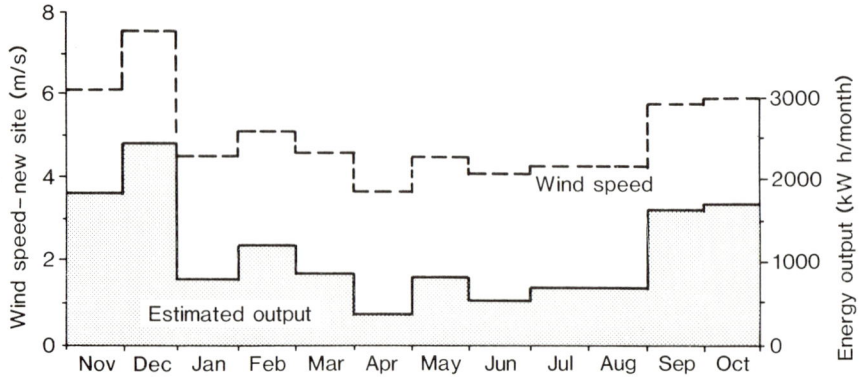

Figure 7.15. The annual wind speed and output from the proposed 10 m diameter wind machine

generator is planned. The new site identified has an annual mean wind speed of 5 m/s at a height of 9 m and the machine's rated power would probably be 8 or 10 kW.

The monthly energy output from the new 10-m diameter machine has been estimated (Fig. 7.15) using the same method as for the Elektro in Fig. 7.14. It has been established, using data measured on the site, that for quite short periods of a few weeks the wind-speed duration curve fits the standard Weibull function:

$$\phi = \exp\left[-\left(\frac{V}{k\bar{V}}\right)^{\beta} \right]$$

where ϕ is the probability of the wind speed exceeding V, $\bar{V}$ is the mean windspeed for the period and β and k are constants.

$\beta = 2$ and $k = 1.2$ provide a good fit to our data. Using this expression, a computer program applies machine characteristics to the wind data, incorporating the cut-in wind speed, furling speed, efficiency at rated power and an adjustment for reduced efficiency at low power levels. The energy output is totalled for the period concerned and is shown on a monthly basis in Figs 7.14 and 7.15.

For the proposed 10 m machine, power output is represented by:

for $V < V$ cut in $P = 0$

for $7.8V^3 < \dfrac{PR}{4}$ $P = 0.5 \times 7.8V^3$

for $\dfrac{PR}{4} < 7.8V^3 < \dfrac{PR}{2}$ $P = 0.75 \times 7.8V^3$

for $7.8V^3 > \dfrac{PR}{2}$ $P = 7.8V^3$

for $7.8V^3 > PR$ $\qquad P = PR$

for $V > V$ furl $\qquad P = 0$

where V is wind velocity in m/s and PR is Rated Power (watts). In Fig. 7.15 cut-in speed is taken to be 3 m/s and PR to be 10 kW. The annual output is estimated to be approximately 13 600 kW h based on fairly pessimistic efficiency assumptions. Improving the low power efficiency and the rotor performance could raise this to around 20 000 kW h.

DISCUSSION OF CHAPTER 7

P. Mudd (Yarsley Technical Centre). Dr Todd's figures for the collector losses in their solar-heated exhibition building span between 60% and 88%. Are these losses typical and if not, what solar collector losses would be expected, what are the main reasons for these losses (i.e. is it materials, construction or design) and what suggestions for improvement can be made to reduce those losses?

Dr R. W. Todd (National Centre for Alternative Technology). Such collector efficiencies are not untypical of simple, flat-plate collectors. The typical annual average efficiency for a simple, flat-plate collector is probably about 30% in a fairly low-temperature application and a good deal less than that if it is being used for applications requiring rather higher temperatures. The loss is almost entirely heat which is transmitted back through the cover plate – a mixture of radiation and convection. It can be reduced by multiple glazing or by selective surfaces. The losses in the installation in the exhibition building relate to a trickle collector which has an additional heat loss due to evaporation of the water and condensation on the cover plate, which is particularly bad at higher temperatures. In the July week shown, where the collectors are working with a fairly high temperature store, the efficiency is probably rather lower than would be achieved with a normal closed collector.

Dr J. C. McVeigh (Brighton Polytechnic). I think the trickle-type collector is not going to be used very much more. Even as far back as 1975 I strongly advised against the use of the trickle collector for the Granada Television 'House for the Future' project* but it made extremely good television to have people clambering around and was something a do-it-yourself person would understand. As a result the trickle roof went in and was seen by millions. The unfortunate result was that general solar-space heating was criticized because an extremely poor system had been used which to the informed was clearly not the best way to do it.

Dr R. W. Todd. Much time and effort was spent on the choice of collector and finally the cheapness of the trickle collector swung the decision in its favour but the extent to which the efficiency was reduced, particularly at high temperatures, was not then appreciated. While accepting average collector efficiencies are at the moment low, other work, for example, the work by Phillips described in Chapter 3, on much more advanced collectors would give probable efficiencies around 50%.

* Trueman B. and Wilson, D. R. (1978) Experience from the 'House for the Future' project of Granada TV, in *Ambient Energy and Building Design* (ed. J. E. Randell), The Construction Press, Harlow.

Dr S. J. Wozniak (Building Research Establishment). Many years ago it was a popular claim that 45% or 50% collection efficiency could be realized in typical domestic hot-water systems. We now know that a realistic figure for good flat-plate collectors is more like 15–30%. The efficiencies quoted by Dr Todd for trickle collectors are, therefore, not pessimistic when assessed against what we now know about well-designed systems.

J. Russell (Newcastle-upon-Tyne Polytechnic). The Centre for Alternative Technology have published their results for a wide range of experiments utilizing wind, water and solar energy sources and paint a somewhat depressing picture since their figures show that of the total energy requirement for the site only 0.3%, 3% and 4% was met by wind, water and solar sources respectively. These extremely poor results must surely require a change in experimental direction. Where do the Centre go from here?

Dr R. W. Todd. Our original aim when the Centre started seven years ago was to gradually develop towards as high a proportion of ambient energy use as possible. There are probably two reasons why we have not got further than we have. One is the state of the technology and our understanding of the technology. The field is not as far ahead as we perhaps thought and in some cases practical systems which work well are not yet available. The other main restriction is capital cost. Most ambient energy technologies, rather like nuclear power, have low running cost and very high initial capital cost. There are also obviously some natural resource limitations on what can be done. The water fraction shown in the paper can probably be increased by improved efficiency but there is a finite limit to the water available on our particular site. The contributions shown in Fig. 7.13 we hope to achieve over the next couple of years and will result in a contribution from ambient energy of 39% of the total useful energy, more than doubling the present level. Extension from this point to 100%, apart from capital costs, also requires many technical problems yet to be solved, particularly on energy storage and in improving the reliability of systems. For example although most of the wind machines we use are commercial products, they are far from being a power source that can be just bought, erected, switched on and forgotten about. Reliable machines are starting to appear, for example, in the USA, but even there the Government Wind Turbine testing centre at Rocky Flats have had very similar experiences to our own and they have produced some well-documented reports which show that most machines have actually failed, often quite catastrophically. The energy is there, the technology is half way there but we are not yet at a buy one and switch it on stage.

8

Energy Strategies and Control in a Hypermarket

Julian Keable

A BRIEF DESCRIPTION OF A HYPERMARKET AS A SEPARATE BUILDING TYPE

Hypermarkets, as developed on the Continent and more recently in the UK, are more than just large supermarkets. Defined as having more than 4647 m² (50 000 square feet) selling space, they typically have a gross area of twice this amount, since they represent a completely autonomous unit in many respects, despite being linked with larger retailing groups. Thus, in addition to the sales area they will have a substantial amount of office space including the whole hierarchy from Managing Director to Assistant Buyers, Accountants, Security Staff and so on. The hypermarket is also a production unit, in that the fresh food is prepared on the premises, and a full-fledged bakery, butchery, and delicatessen unit will typically be included.

Thus, a hypermarket has very widely varying design requirements. Low- and high-temperature cold stores, a large number of cold cabinets in the store itself; air conditioning for the sales hall and the office space; heating and ventilating in other areas will normally be required.

ENERGY USE PATTERN

A hypermarket will normally be occupied to a varying extent throughout the year. During the hours of opening, typically between 9.00 a.m. and 8.00 p.m. on Tuesday to Saturday there may be some 300 to 500 staff and anything from 1000 to 3000 customers inside the building at any one time.

It is a feature of such buildings that restocking of shelves can take place outside opening hours, since large stocks are held in-house in the adjacent warehouses. A smaller workforce will be in occupation at those times. At other times, security staff and maintenance men will be at work.

It is clear from this that the level of incidental gains that will be available in such a building must vary through very wide limits. The heat gain from occupation by people may be inferred; the heat gain from artificial lighting will also add substantially, since fairly high lighting levels are normal in such stores. Heat gains from equipment will be lighter, though significant in some areas such as in food preparation areas, and the restaurant kitchen (a restaurant for staff and public is usually included) (Fig. 8.1).

Indirect gains are, of course, also available from the extensive refrigeration equipment, and some measure of heat recovery from this source has become fairly common. Typically, reject heat has been routed to the warehouse areas so as to act in heater banks there when required.

THE 'TRADITIONAL' ENERGY SOLUTIONS FOR A HYPERMARKET

This building type, often thought of as having originated in the USA, but in its present form having been largely developed in France and elsewhere on the Continent, was a product of those years, now seemingly distant, when energy was cheap. The typical solution, therefore, solved each problem as it arose by the application of a separate, convenient form of energy.

Air conditioning employed pre-fabricated roof mounted units of direct expansion type, and originally using electric resistance coils for top-up heating. More recently the typical solution for space heating in the sales hall and elsewhere, was to include heating coils in the roof-mounted units, for low-pressure, hot-water supply from an oil or gas-fired boiler.

Fluorescent electric lighting was the norm, though some examples of discharge-tube lighting have been seen; the refrigeration is, of course, also electrically powered. Gas was used for cooking, both the fresh food for sale in the shop and for the restaurant kitchen. Hot water, principally used in the preparation of food and for washing-up, would come from the central boiler house.

Because of the heavy dependence on electricity, it became standard practice to install an emergency generator, capable of supporting the essential refrigeration load and some lighting and other essential loads.

No attempt was made to link the wide variety of energy-using equipment in any way, and only recently has some attempt been made to avoid heavy penalties for 'maximum demand' tariffs.

Figure 8.1. View of food preparation area from sales hall: butchery

THE NEW CHALLENGES

Rapidly rising energy costs are now a byword, and it is even part of government policy that there should continue to be a relative inflation between general expenditure and energy expenditure for some time to come. In addition, there has developed recently the 'gas embargo' whereby only 25 000 therms are available for connection to any particular new building from the Gas Board. Further factors affecting energy strategy include the uncertainties of deliveries (and therefore the desirability of being able to operate for short periods from an on-site reservoir), and likewise the desirability of flexibility of energy strategy wherever this is possible.

It is important to emphasize, however, that the *raison d'être* of a hypermarket, namely the efficient marketing of products at a low price, precludes the use of design solutions which are not cost effective.

One further factor concerns the new building regulations including the insulation standards required of new buildings. This needs some care in handling in the case of a hypermarket, since the high internal gains lead to a low balance point, and hence the need for air conditioning. Without due care, insulation could lead to an increase in energy use if electricity were to be applied throughout a longer period of the year as a result of the prevention of heat loss from the fabric!

THE HAVANT HYPERMARKET

The building now to be described is a new building, recently developed by the Portsea Island Mutual Co-operative Society, an independent co-op operating in the south east (Figs 8.2 and 8.3). Architecturally designed by Mr Marcus Pegg, the Society's Staff Architect, the building is the first of its kind to be undertaken by this Society. Although conforming in many respects to the general lines pioneered by Carrefour in this country, the design solutions differ in many ways and in detail from that prototype. It is also perhaps the first hypermarket to have been built under the new building regulations, and certainly the first to be built after the introduction of the 'gas embargo'.

Helix was appointed sufficiently early that it was possible to incorporate the design strategies proposed into the architectural design as it developed, so as to achieve the maximum integration. These strategies include:

(a) Energy conservation by:
  (i) direct heat reclaim within the occupied space
  (ii) indirect heat reclaim from the refrigeration circuits.
(b) The use of 'free cooling' from the ambient air whenever this is possible (Fig. 8.4).
(c) Back-up space heating to be provided by electrically driven, heat-pump units on the roof (Fig. 8.5).

Figure 8.2. Aerial view during construction showing roof-mounted units

Figure 8.3. Hypermarket at Havant – an external view

Figure 8.4. Detailed internal view of air handling units in sales hall

Figure 8.5. Roof-mounted air handling units and heat pumps

(d) Alternative back-up heating using propane gas through an indirect, gas-fired, warm-air heater within other roof-mounted units.

(e) The use of computerized monitoring and control, encompassing maximum demand control but more importantly the interaction between the demands from different parts of the building and possible sources of reclaimed energy (Figs 8.6 and 8.7).

I now turn to each of these matters and consider them in more detail. Before doing so, however, it might be worth recalling the nature of the air to air heat pump.

The operating principles of a heat pump are becoming relatively well known, but attention is not always paid to the source from which the heat pump will gain

Figure 8.6. Control panel with door open

Figure 8.7. Energy control unit with command module in foreground

the additional energy in the cycle. Where the heat pump is using waste heat in an industrial process for example it cannot, in any sense, be described as a solar device. Where, however, the source is ambient air, then the source of additional energy is clearly solar. Air, the temperature of which is held above absolute zero by the action of the earth's mass (the earth and ocean acting as solar energy storage vessels), transfers that stored heat throughout those periods when direct solar energy is not available, including of course the night time. Thus, a heat pump, an essentially hybrid piece of equipment, operating at a coefficient performance in heating of 3 to 1, could be described as a device delivering 66% of its load from an indirect solar collector while 33% of the load comes from (in this case) electricity.

Heat load

Insulation
Normally, and especially in a building of this type with high internal gains, arising from occupation and lighting, compliance with part FF of the Building Regulations provides an adequate standard. Natural lighting from north lights might be considered where acceptable, but is not suitable for a retail environment.

Loss of heat through envelope
Up to 50% of heat losses result from air changes. The main loss arises from door openings in the external walls. Therefore necessary external openings are protected by strip curtains or lobbies as appropriate. Significant losses can also arise from 'leaks' in the envelope. Great care has therefore been exercised to ensure that joints in and between materials are designed to minimize leakage.

Heat emission
The height of the building produces a natural stack effect. The coldest air tends to rest at floor level whilst the hottest is at high level where it is not required and is wasted. To counteract this effect, return air is drawn from high level and redistributed at floor level. This arrangement can reduce fuel consumption on heating by 10–15%. Heat loss through the roof is also reduced.

Control of air changes
In order to achieve satisfactory ventilation characteristics, some 6 l/sec needs to be introduced into the building per occupant. This can account for some 20–30% total heat use. It has been common practice to design a system based on maximum occupancy with the fresh-air requirement based on this criterion.

In this case the system takes account of occupancy. For example, no fresh air is brought into the building during the night nor during the morning warm up period. Maximum air for ventilation occurs only when the building is fully occupied. Furthermore, the system is controlled so that the maximum use is made of fresh air for cooling, thus minimizing the cooling load.

Energy control system
In order to control effectively the complex electrical systems a microprocessor is required, which among other things:

(a) Monitors and controls maximum demand levels by selective load shedding, and cycling of plant.
(b) Introduces a sophisticated time control system on all plant. The client can and does reprogramme each function in accordance with the actual use pattern of the building, so that a daily sequence of lights, fans, pumps and so on is adjusted in the light of experience (Fig. 8.8).
(c) Optimizes the use of the standby generation system.

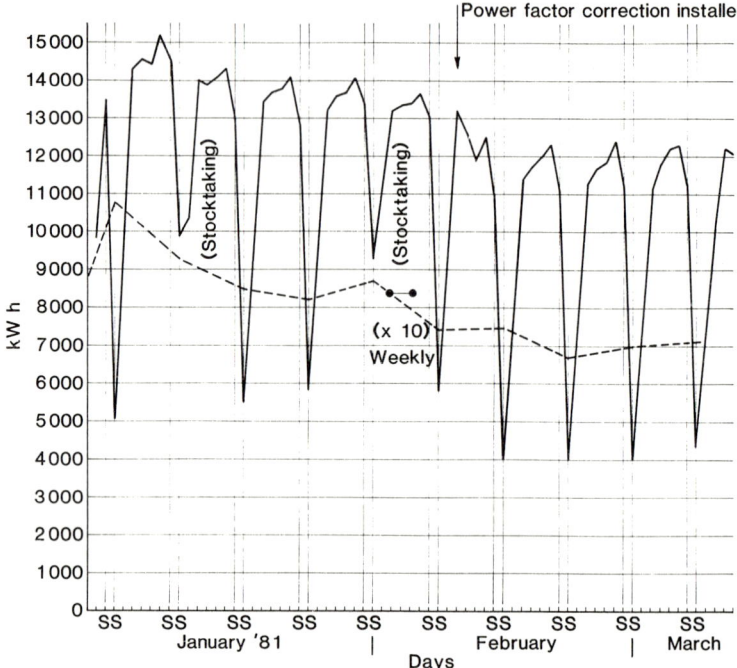

Figure 8.8. Reduction of energy use during a period of increasing sales

The system has other functions which would warrant its own discussion paper. The saving in electricity charges will pay for the equipment in under two years.

Heat recovery

Refrigeration plant
In this building with its high refrigeration requirements for food preservation, some 60% of the *maximum* heat requirement for space heating and hot water can be recovered from this source. Above an ambient of 4°C, no other heat source is required for space heating in the sales hall. The heat intended for space heating is applied direct through the air conditioning units by way of condensers in the units. Hot water is recovered by de-superheating the refrigeration machinery to obtain the highest possible temperatures. The heated water is passed to a conventional hot-water cylinder to present the boiler with water at above mains temperature. Control is not necessary.

Efficient use of reduced energy requirement

Heat pumps
Most of the remaining heat requirement is met by eight roof-mounted Lennox

CHA packaged heat pumps. These provide heat more cheaply than an oil or LPG fuel system.

Some machines perform a cooling role during the summer. Unfortunately, heat pumps become less efficient as ambient temperature drops and their heat output falls. This is, of course, opposite to the building heat requirements which rise as ambient temperatures fall. In America it has been common to make up the shortfall in cold weather by using direct-acting electric heaters. These are expensive to operate and bivalent (two-fuel) systems are now being used.

In this building, therefore, LPG fuel units 'top up' the heat pump system in cold weather. Similarly this system supplements the 30% of hot water provided by the refrigeration system.

BUILDING DESIGN CONSTRAINTS

It can be seen that the integration of these systems within the building has been possible without altering the volume or constraining the design in significant ways. The roof-mounted units have tended to be larger in number and smaller in size; space requirements for the heat recovery from the refrigeration compressors are minimal, and other roof-mounted equipment uses no useful volume.

The same remarks would apply to other strategies which we considered but were not adopted, and are described in the following sections.

Strategies considered but not adopted

Underfloor heating
Underfloor heating of the 'STRAMAX' type was considered since low-grade hot water at 35°C can be used which matches well the temperature of refrigeration reject heat. Also the heat is at the right level in the building and is cheap to circulate. The system was ruled out on cost and lengthy installation time.

Sprinkler tank for storing heat
The use of a large volume of water for storing reject heat appeared to have attractions. The problem is that a low-pressure, hot-water system is required to pass the water to the heating units. This is very expensive since the temperature is low. The alternative described earlier is achieved at one-third of the cost.

Total energy
The effective application of the concept of total energy depends on there being a match between electricity demand and heat requirement. In this building the electricity demand for lighting and refrigeration is high and therefore the waste heat generated would be well in excess of that required. The economical solution is that adopted of using the heat recovered from the refrigeration process as a form of 'heat pump'.

Comment

With the advent of the direct-fuelled engine heat pump, all these strategies will have to be re-thought once again.

Approximate technical data

(a) Total building heat loss at 1°C is calculated at 400 kW.
(b) The balance point of the building allowing for heat recovery, lighting and average occupancy is +4°C.
(c) The refrigeration plant has an installed capacity of 210 kW.
(d) The heat produced by internal lighting is 150 kW.
(e) The average heat from refrigeration to heat water is 30 kW.

DISCUSSION OF CHAPTER 8

Dr S. J. Wozniak (Building Research Establishment). To what extent is it still considered to be necessary to design a heating and cooling system for a commercial type of building, the aim of which is to keep the building within a narrow range of temperature during all occupied hours? Additionally, to what extent is it thought that large savings in both capital and running costs might be possible by widening the band of realized internal temperature?

J. Keable (HELIX Multiprofessional Services). There is no doubt that in the hypermarket described there is a lot of flexibility in the band of temperature which is acceptable because the space is so large. The volume of air per user is also large so that fresh-air supply is based on human need. The air flow and the fresh air heat load are small by comparison with the totality. The conditions that are paramount in a hypermarket concern is food preservation, not people, hence the huge emphasis on refrigeration. A private view rather than a general answer to the question is that in buildings where the possibility exists of soaking up ambient gains from any source there is greater possibility of allowing a spread of temperature yet still remain within a comfort band. It is the buildings with very low thermal mass which respond faster to ambient gains and, therefore, to overheating where the problem exists.

Dr S. J. Wozniak. In office buildings for example to what extent are architects and designers still required by clients to install a system which will maintain a constant internal temperature? To what extent is it now realized that by allowing the internal conditions in the building to drift to some extent a considerable fuel saving might be achieved for very little effort?

J. Keable. Clients vary enormously in their requirements and often it seems they have much more regard to inanimate objects, e.g. food, or computers, with exacting thermal requirements. There is, however, a readiness to listen to the arguments and we certainly stress that if it is possible to accept a drift of say 5 K (2½ K either side of a design condition) considerable saving can be made.

Dr S. J. Wozniak. These days the air-conditioning system for a computer can cost more than the computer itself. Large savings in capital and running costs may be possible simply by not air-conditioning the complete room but by positioning the cooling fans within the computer more sensibly. A wide range of room temperatures can be acceptable.

F. E. Nicklin (Ryder & Yates & Partners). To return to the question of the volume of the hypermarket. Could Mr Keable say how a decision was made on the height of the hypermarket, how was the space insulated? What are the *U*-values? What is the influence of that very high volume on the energy balance of the building as a whole? I have a hunch that such high volumes may provide a great deal of help in energy conscious design.

J. Keable. There are a number of places in the building where a certain height is needed. Parts of the building are on two floors; in other parts elements which don't take up ground floor space like water-stores for the sprinkler system; in yet other parts there is high bay stacking for goods. This establishes a height requirement and in the interests of economy the roof is taken over the whole space making it into a box.

It would be possible to change the heights at different places if the aim was to minimize the volume. We believe and have established by calculation that the volume itself is a useful buffer, provided any stratification effects are overcome. I mentioned in the chapter that the idea of heating underneath the floor was considered: this is the ideal place because it is where the heat is needed. The floor provides a large surface there so very-low-temperature heating (35°C distribution for example) could be used – this is a good match for heat pumps and heat recovery. It was not done partly because of cost but mainly because the building programme required very rapid building.

J. Field (Solar Energy Developments). Was an overcost and energy-saving exercise carried out for the hypermarket. If so, what were the results?

J. Keable. Certain equipment was looked at in this way. It was necessary to justify the microprocessor for example as a straight extra to cost. The Portsea Island Mutual Co-operative Society is an independent body who had a specialist adviser from the Co-operative Wholesale Society. Paybacks in the order of up to three years were generally considered to be reasonable although not all equipment met this criterion. It is difficult sometimes to justify a particular item of equipment whereas a range of equipment taken as a whole can be justified. The overcost as a whole is of the order of a modest £80 000–£90 000 in comparison to the cost of the project – £1.3 million for the services contract alone. The building had to comply with part FF of the Building Regulations and the U-values used are those laid down in those regulations, i.e. higher insulation standards were not used. For a building with a very low balance point, there is at times difficulty in justifying insulation at all, because if the requirement is to lose heat insulation is disadvantageous. The cold rooms run on a 24-hour basis and have to reject heat. In cold weather this heat is rejected straight into the sales hall and the large air volume allows this to be done throughout the 24-hour period. That means that the question of insulation does become worthwhile at night in these circumstances and the usual things like optimum start control becomes out of place.

A. R. Tanner (Wessex Regional Health Authority). Fluorescent lighting appeared to be used throughout. Was there any scope for more efficient forms of lighting, particularly in view of the high ceilings because of the saving in the costs of relamping.

Mr Keable said there was little cooling. This seems surprising in view of the internal gains from occupants and lights which would indicate little heating rather than little cooling. Was any conscious account taken of the stack effect to reduce cooling load and did the very long runs of refrigerant piping to the heat pumps on the roof present any problems?

J. Keable. Certainly thought was given to other forms of lighting but very few shops of this kind have used discharge tubes. A problem with the discharge tube is that if power is cut, on restoration a warm up period occurs before the lights come on again. The inherent security problem is obvious and so is the great reticence on the part of the traders to use such lighting. High efficiency lamps were installed – the costs again needed justification, but seemed worthwhile, not only because of energy saving but because of the extra costs of relamping.

Of course cooling is needed, but ambient air is for the most part quite sufficient for space cooling. On any particular day the need for a warm up in the morning may arise (using reclaimed heat) while later in the day there is frequently a need for cooling, usually by increasing fresh-air volumes.

The stack effect is used for cooling by extracting air at high level, reinjecting fresh-air at low level. I would not agree the refrigerant lines are exceptionally long, although greater than if heat recovery was not used. The refrigeration industry seems very used to, and much more adept at, running lines with good insulation around them than is the heating industry – this is perhaps an indicator of the best way to move heat from one place to another.

J. Harrington-Lynn (Department of the Environment). Section A5 of the *CIBS Guide* (1979) contains a theoretical model, which makes an allowance for the effects of large volumes on ventilation rates. Comment has been made about the effect of Building Regulations on the energy design of buildings. Part FF of the Building Regulations does not *require* such buildings to be insulated. The deemed-to-satisfy provisions stipulate levels of insulation, but these are only the simple solutions. It is open for a designer, by calculation, to demonstrate his building is energy conserving and ignore the deemed-to-satisfy requirements.

9

Energy Conservation in Building Design

P. O'Sullivan and P. J. Jonas

INTRODUCTION

Over recent years designers have been making serious attempts to conserve energy in our buildings. The question, therefore, this chapter attempts to consider is: What role have ambient energy gains played and can play in reducing the energy usage in our buildings? The traditional view is, of course, that ambient gains are useful and can save significant amounts of energy. It seems appropriate, therefore, to review where we have got to, what we know and what our hopes and aspirations are for the future in this particular aspect of energy conservation when ETSU are beginning a new passive solar energy programme (which is of course a large part of ambient gains) and when the HDD are considering the results of their 'Better Insulated Houses' programme.

Ambient energy is often thought of as including the energy derived from solar gains, occupants, lighting and other miscellaneous gains. These gains are both radiant and convective and can be estimated in a variety of ways. Namely:

(a) *Solar gains:* The total solar gains are either measured, or calculated, using solar geometry on an hourly or daily basis and from these total gains on the horizontal, the gains on each and every facade, are predicted. In the UK we use either the ASHRAE [1] (Solar Geometry and Cloud Cover), the CIBS [2] and/or the Berlagge [3] methods of prediction. These methods produce typical results of:
  (i) A semi-detached house – 4 kW h per day in January and 13 kW h per day in March/October.
  (ii) A typical school (1500 m² and 300 pupils) – 100 kW h per day in January and 250 kW h per day in March/October.

(b) *Occupants:* These are estimated from the results of long-term, large-scale, physiological measurement programmes. The DES suggest 70 W per person [4], the CIBS 80 W per person (at 22°C)[2], and Siviour [5] 80 W per adult and 60 W per child. Again, these results suggest for a typical house, 4–6 kW h per day gain and for a school a 200 kW h per day gain.

(c) *Lighting gains:* These are usually based on the connected electrical load and then broken down into that energy which produces light and that which produces heat. The balance depends, of course, on the light source and light fitting, but again typical results for houses, 1–2 kW h per day [5], for schools 12.6 kW h per square metre per annum [4] and for offices 35 kW h per square metre per annum [6].

(d) *Miscellaneous gains:* Includes cooking (houses), small power in offices, etc. These are again based on the connected load and the efficiency of the appliances and result in gains of, for a house, 10–15 kW h per day [5], for a school, 3 kW h per square metre per annum [4] and for offices, up to 6 kW h per square metre per annum [6].

To assess what contribution such ambient gains make to the space heating load of a building, it is usual to attempt some predictive calculations. In such calculations it is normal to sum together the individual ambient gains and to subtract this total from the total heating load. This calculation can be carried out either on an hourly, daily, or on a seasonal basis, depending on the method of calculation used. The more sophisticated the method of calculation/prediction, the smaller the time step used. For example, if a finite difference model was to be used to model the energy requirements of a typical terraced house, the data file to describe the ambient gains would be as in Table 9.1. From such a file the thermal

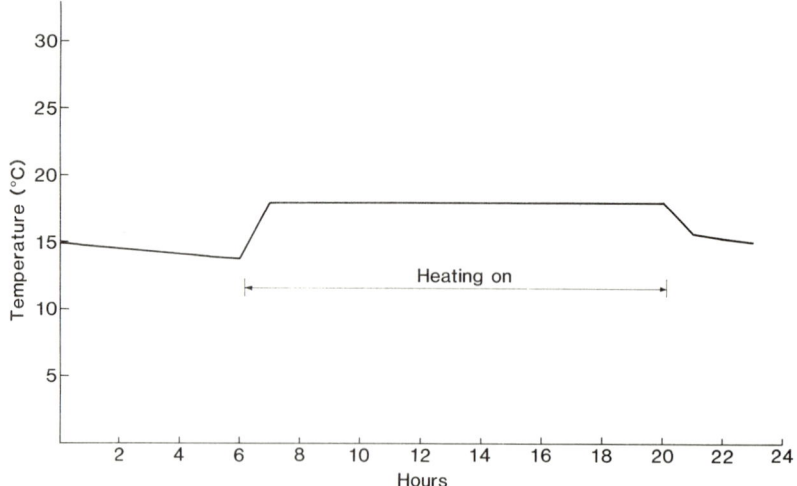

Figure 9.1. A typical temperature profile for one day generated by the finite difference model. The inside thermostat is set at 18°C

Table 9.1 Input data file for finite difference model listing ambient gains

Hours	People (no.)	Lights (kW)	Misc. (kW)	Total horizontal solar (W/m²)	Cloud cover (1/10 s)
1	4	0	0.1	0	0
2	4	0	0.1	0	0
3	4	0	0.1	0	0
4	4	0	0.1	0	0
5	4	0	0.1	0	4
6	4	0	0.1	0	0
7	4	0	0.1	0	0
8	2	0.1	0.3	0	9
9	2	0.1	0.3	33	7
10	2	0.1	0.3	116	5
11	2	0.1	0.3	149	4
12	2	0.1	0.3	179	4
13	2	0.1	0.3	179	7
14	2	0.1	0.3	135	9
15	2	0.1	0.3	86	9
16	2	0.1	0.3	24	9
17	2	0.2	0.3	0	9
18	4	0.2	0.4	0	9
19	4	0.2	0.4	0	9
20	4	0.2	0.4	0	9
21	4	0.2	0.4	0	9
22	4	0.2	0.4	0	1
23	4	0.1	0.1	0	5
24	4	0	0.1	0	1

response of the building could be predicted (Fig. 9.1) as also could be the energy usage (Fig. 9.2). On this basis (Table 9.2), the contribution of the ambient gains can be seen to reach the typical figure of approximately 30% of the total heat requirement [7]. Similar calculations for schools and offices produce results of similar magnitude.

Armed with the 'substantial' values, technical methods of 'obtaining' the gains were derived by engineers which varied from the simple (e.g. letting the radiant heat in through the windows), to the much more complex (e.g. the use of heat pumps in connection with heating systems). Typical solutions have resulted in a combination of passive and active collecting systems which attempt to take advantage of not only the ambient gains that reach the inside of the building and those that are generated there, but also those external to the building [8]. The main thrust of this chapter concerns those gains within the building as the use of the 'external' gains has received considerable attention both in this book and elsewhere.

Table 9.2 The daily and monthly totals output by the finite difference model

	Daily totals January 19 (kW h)	Monthly totals January
Net heat load	48.1	1593.0
Lights	2.1	65.1
Small power	4.9	151.9
Occupants	6.2	193.0
Solar	6.2	116.0
Ambient gains as a percentage of total heat load	30%	33%

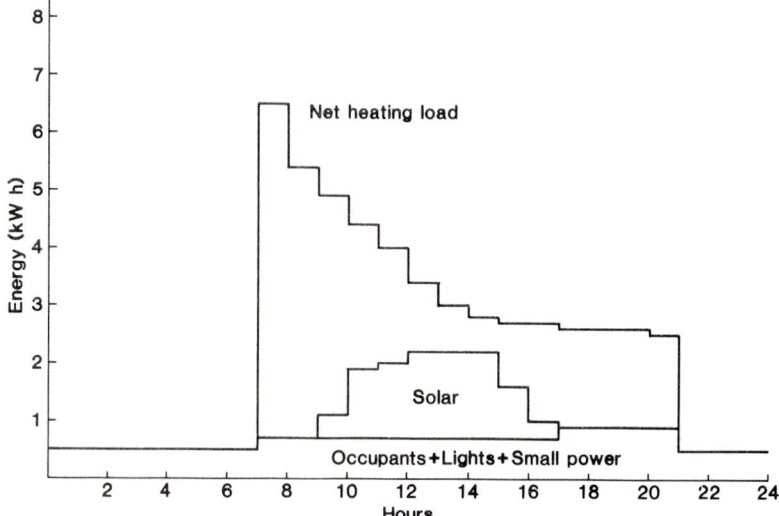

Figure 9.2. The calculated heating load together with the heat energy supplied to the space by the solar, occupancy, lights and small power gains. The calculation method assumes that the ambient gains are fully used

The interesting fact is that both the majority of the methods of prediction of collection are based on an inherent assumption that, 'if one got things right', almost 100% of the ambient gains could be usefully used in some way or other albeit over a rather difficult time scale.

EFFECTS OF ENERGY CONSERVATION

What then are the effects of our recent programmes of energy conservation in buildings on the availability of such gains (i.e. on the traditional wisdom)?

The fabric

In our attempts to conserve energy we have improved the insulation values of our buildings thereby reducing the conductive loss, but increasing the 'retention' of the internal ambient gains. As part of this programme we have reduced the area of glass of our buildings thereby reducing the radiative gains from the exterior. By limiting and controlling the ventilation we have reduced the convective losses from our buildings and by making the building's internal volume both more stable in the summer and thermally more efficient in the winter, we have, in effect, *increased* the potential for using the available ambient gains.

The ambient gains themselves

The availability of externally derived radiant gains has been reduced in direct proportion to the reduction in the area of glass, i.e. by about 50% in our offices and schools and approximately 20–25% in our houses [9].

The lighting levels have also been reduced, reducing the available energy in our offices by approximately one-third, in our schools about 20% and in our housing probably not at all. This latter has been achieved not only by reducing the lighting levels themselves, but also by increasing the efficiency of the light sources.

The energy used for cooking in our commercial and school premises has been reduced by changes in eating habits, but in our houses, not at all.

Occupancy levels, if anything, are increasing in our buildings generally, as also are the incidental (small power) loads.

The ambient gains, therefore, have been reduced in comparison with those buildings designed in the sixties (primarily by the reduction in the lighting and radiation gains), but are still significant.

The system

The boiler sizes of our low-energy buildings have been reduced to allow for a reduced conduction loss. The controls have been improved to ensure that the boiler fires only when required, but the controls sensing the environment of our spaces have hardly changed.

It would seem, therefore, that the design of our buildings to control energy use, increases the potential for making use of our ambient gains, reduced the gains themselves in total, but still inherently assumes that the gains will be useful (by reductions in boiler sizes, etc.) and sensed by the traditional sensors.

EXPERIENCE OF MEASUREMENT

What then have been our experiences in measuring buildings to determine what has happened in practice? It is difficult to evaluate directly the proportion of ambient gains that are used. However, it is possible to evaluate them indirectly from measurements made in buildings.

Housing

From observations made of temperature profiles in occupied houses and also of energy consumption and use profiles, it is possible to determine what happens to the ambient gains. For example, Fig. 9.3 illustrates the temperature profile in the kitchen of a well-insulated house on an hourly basis over one day. The kitchen radiator was fitted with a thermostatically controlled radiator valve (TRV) which was set in the middle of its range. Also shown in Fig. 9.3 is the energy used for cooking and the period of time the heating was on, the kitchen was occupied and when the windows were open. It can be seen that the temperature was not

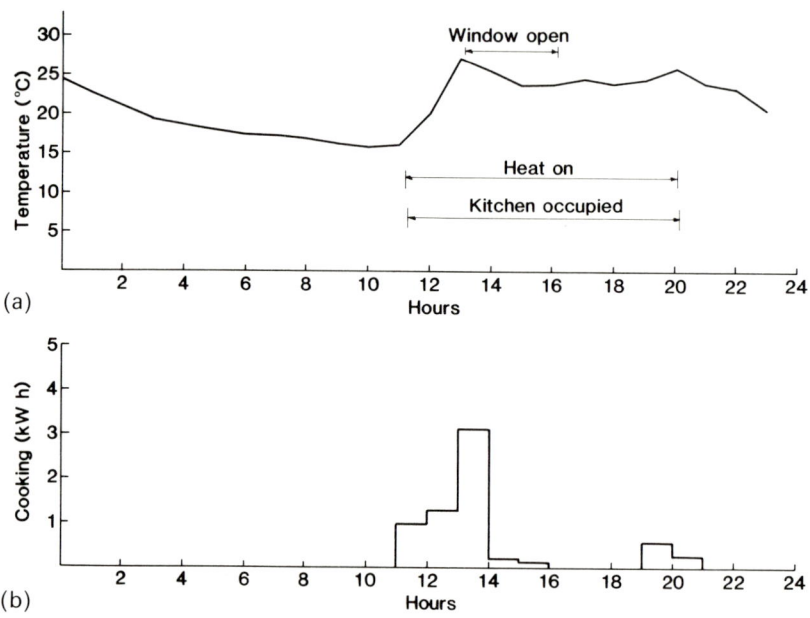

Figure 9.3. (a) Temperature profile of kitchen; (b) energy used for cooking

controlled adequately, it reaching a maximimum of 27°C as a result of the cooking and occupancy. Only when the windows were opened did the temperature come down. In this situation then the ambient gains first produced higher temperatures and afterwards resulted in windows being opened. The question to ask then is: Why doesn't the TRV control the temperature when the ambient gains occur as that is its function? Figure 9.4 shows the temperature gradient from floor to ceiling in the kitchen when the cooker is in use. Although the temperature at the ceiling and mid-levels rises rapidly in response to the

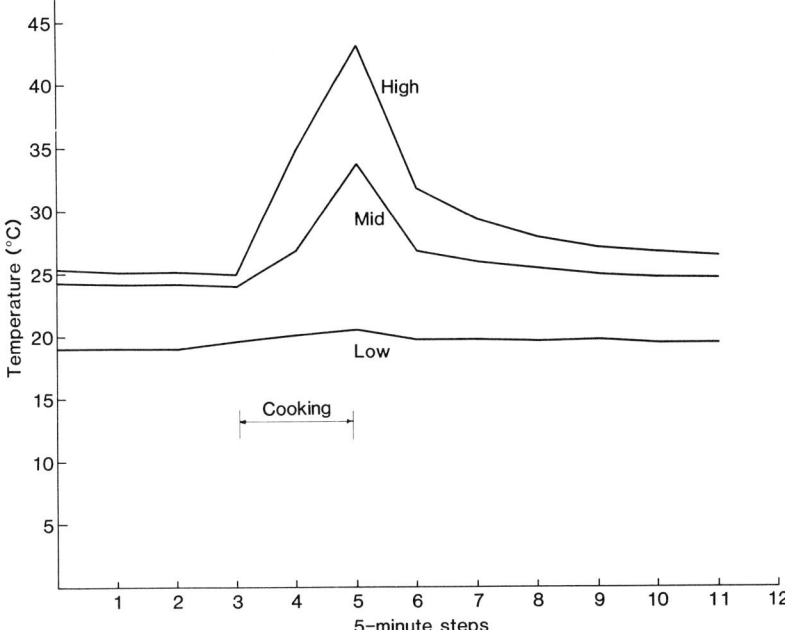

Figure 9.4. Temperature gradient in kitchen (floor to ceiling) during cooking

cooker, the temperature at floor level where the TRV sensor is situated, hardly responds at all. This example illustrates the difficulty of controlling a dynamic situation (both spatially and temporally) necessary for the effective use of ambient gains.

Schools

Here again, the results show the difficulty of controlling a dynamic environment in the presence of ambient gain. Figure 9.5(a) shows the temperature profile for a day when the school is occupied during the heating season with the lights on and when there is a relatively large amount of solar gain. Figure 9.5(b) shows a temperature profile for the same space when the school is unoccupied, there is no lighting load and when there is a low-level of solar gain, but with the heating on. A comparison of the two profiles shows the controls can maintain the required temperature when there is no ambient gain (Fig. 9.5(b)), but when the ambient gains are present (Fig. 9.5(a)) there is a build-up in temperature throughout the day!

These examples illustrate the difficulties involved in making full use of ambient gains. From measurements made in houses and schools we can again see the significance of ambient gains and their potential for contributing towards space heating. Table 9.3 shows measurements made in houses and schools. The data for

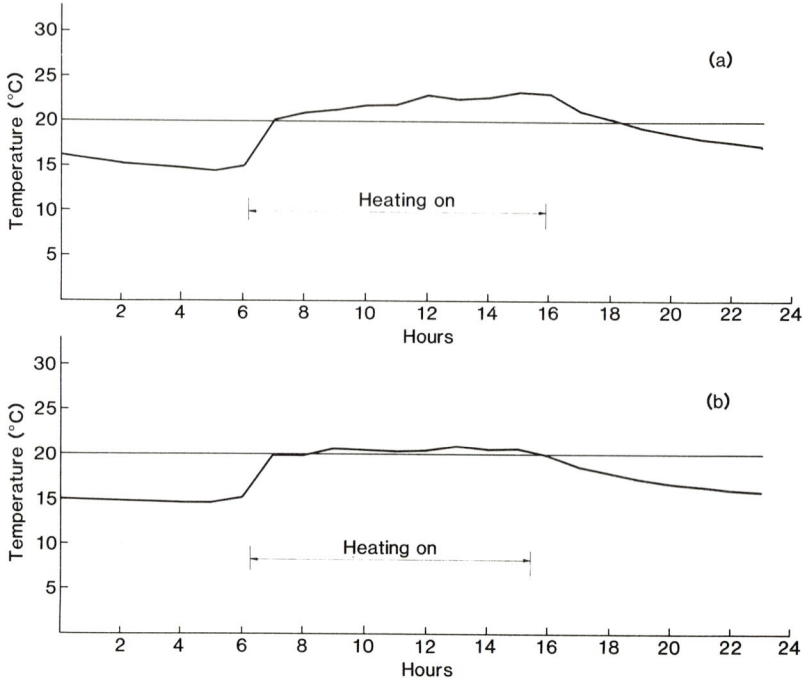

Figure 9.5 (a) Temperature profile for a school over 24 hours. The school is occupied with the lights on with a relatively large amount of solar gain (during the heating season); (b) temperature profile for the same school over 24 hours. The school is unoccupied, with no lights, and a low level of solar gain (during the heating season)

Table 9.3

	Houses (design day heat loss ~70 kW h/day)	Schools (300 people ~1500 m^2) (design day heat loss 2000 kW h/day)
Occupants	4–6 kW h/day	150–200 kW h/day
Lights	1–2 kW h/day	240 kW h/day
Miscellaneous	2–6 kW h/day	–
Cooking	2–6 kW h/day	–
Solar*	4–15 kW h/day	90–300 kW h/day (Dec–Mar/Oct)

* Estimated from measured total horizontal.

houses was obtained from measurements made over a number of houses and it is interesting to note the variation for the individual sources of gains. What is also perhaps of interest, is that houses with high levels of ambient gain are usually houses with relatively higher heating-energy consumptions.

CONCLUSIONS

From the experience gained in measurement projects a set of general conclusions and observations can be outlined concerning the use of ambient gains in low-energy buildings.

(a) To be of use ambient gains must occur either when and where they are needed, or be able to be transferred to the place and/or time of need. Where possible spaces and the use of spaces must be designed and organized to coincide with the occurrence of ambient gains. To some extent this happens naturally, occupants often causing the gains, e.g. cooking. Open-plan design may prove more favourable in exploiting ambient gains, making them easier to distribute and less likelihood of them resulting in overheating.

If the gains are to be used passively, thermal syphoning and buffer zones may prove necessary to distribute the gains and stabilize the thermal environment. It may prove necessary in many cases to use active systems to collect and distribute the gains.

The serious use of ambient gains will have an effect on system control. Decision must be made such as, should a building with a high occupancy level be warm on entry or should the people, lights etc. warm it up. If the trend towards localized lighting systems continues, lighting gains will be less but they will also be more difficult to control and must therefore be considered differently in system design.

(b) The ambient gains must be of a suitable magnitude, duration and form. From our observations of gains in practice, they can be divided into two forms, steady gains and pulsive gains.

The steady gains, e.g. lights, occupants, can usually be specifically designed for and should be able to be adequately controlled and integrated into the system design. The pulsive gains (especially on a domestic level) are more difficult to control and are often of less use. Often they have to be restricted (solar) or in some cases 'got rid of' (cooking). If pulsive gains are to be used, they must be quickly sensed so that the system is able to respond to them in time.

The form of ambient gains is also important regarding their potential for use, i.e. that radiative/convective split and moisture/pollution levels.

For example there is a problem of high radiant temperature and glare in spaces where there are direct solar gains. There is also the problem of high moisture levels in heat gains from cooking. On an industrial level there is the problem of distributing warm air in situations where pollutants are present.

If such ambient gains are to be used, methods must be investigated for changing their form, e.g. getting rid of moisture, filtering out pollutants. The question then arises as to whether it is worth it?

(c) Ambient gains must be sensed efficiently and the systems/buildings must be able to respond to them. If ambient gains are to reduce the amount of energy the heating system uses they must be sensed effectively in order that the system is able to respond to them.

In many situations the sensing device is unsuitable – either it is measuring the wrong parameter or it is badly sited with respect to the occurrence of gains. This is especially a problem for cellular buildings. With the growing application of microprocessors it might prove more effective to detect rates of change of various parameters rather than absolute values.

What is apparent from field study especially in houses is the need for more local control over heat emitters so that users can actively respond to gains. So that when ambient gains occur they are not negated by an inefficient system, the system should be able to operate efficiently at part load, e.g. houses – lightweight boilers, larger buildings – modular systems. Boilers must be sized to consider gains and heat emitters should be able to allow for gains to be used, i.e. they should have a fast response.

THINKING OF AMBIENT GAINS IN TERMS OF FUTURE DESIGN

From these 'rules' some implications for future design can be listed:

(a) Ambient gains, can be divided into two types:
   (i) Pulsive – e.g. solar, cooking.
   (ii) Steady – e.g. lights, occupancy.
(b) Buildings, or parts of buildings, can be broadly divided into two types:
   (i) those that are occupied during the day (e.g offices, schools) and
   (ii) those that are unoccupied (e.g. houses, hotels).
(c) Where unoccupied buildings are concerned, both types of gains, i.e. pulsive and steady, are of direct use (over-heating not being so critical).
(d) Where occupied buildings are concerned:
   (i) the steady gains should be designed for (although measurements have shown that at present there seems to be a control problem)
   (ii) the pulsive gains should in many cases be restricted and/or be used indirectly.
(e) Indirect use of pulsive gains can take the following forms:
   (i) active systems, e.g. heat pumps, solar collectors
   (ii) passive systems
   (iii) used to effect a better outside environment.
(d) Active systems – maintenance and utilization problems are slowing the process down.

(e) Passive systems – the thinking about passive systems in the UK is still in early stages of development. There is the arrangement of spaces to consider, e.g. the design and use of unoccupied spaces, bedrooms, etc. for use as collectors/buffer zones.

(f) In order to use gains outside the building, we must develop our thinking about properties of buildings for example the use of courtyards, and perhaps use these outside spaces as a heat source for heat pumps.

To initiate this development of thinking about the future of buildings and the use of ambient gains we must consider the following areas:

(a) A better knowledge of operation and use patterns. This is a key area of research and one in which we are beginning to make some progress.

(b) The design of the inside and outside fabric to effectively control the inside and outside environment.

(c) The design of spaces to match with the occurrence of gains and use patterns.

(d) The design of system and controls. How much can we reduce the size of our boiler and how should we control? Should we, for example, use a control which anticipates ambient gains and as a result switches off the heating system. The system comes back on either if the gains anticipated do not occur or after the gains have occurred.

(e) New hardware: we must consider the use of ambient gains in the development of new hardware and bear in mind that in the use of ambient gains we may be conserving 'cheap energy' but we are increasing our dependence of electricity.

(f) Natural control of the environment. If more sophisticated controls are to be used in non-air-conditioned, forced-ventilated buildings, then we must avoid the dangers of developing these controls around the wrong assumptions of use patterns, etc. In other words we should not let technology take off before we understand where we want it to go and we must order our schedule accordingly.

REFERENCES

1. American Society for Heating, Refrigeration and Air Conditioning (1975) *ASHRAE Handbook*, ASHRAE, Atlanta, Georgia, USA.
2. Chartered Institute of Building Services (1970) *IHVE Guide Book*, Book A, CIBS, Balham, London.
3. Basnett, P. (1978) *Curves for Determining Solar Radiation Incident on Vertical Surfaces*, ECRC/R1199, Electricity Council Research Centre, Chester.
4. Department of Education and Science (1979) *Guidelines for Environmental Design and Fuel Conservation in Educational Buildings*, HMSO, London.
5. Siviour, J. B. (1977) *Calculating Solar Heating and Free Heat and their Contribution to Space Heating in Buildings*, CIB Steering Group S17, 'Heating and Climatization', Holzkirchen, September.

6. Chartered Institute of Building Services (1977) *The IES Code for Interior Lighting*, CIBS, Balham, London.
7. Department of the Environment (1980) *Better Insulated Houses, Abertridwr, Monitoring of Domestic Central Heating Systems*, unpublished report.
8. Department of Energy (1980) *Solar Heating Programme – Projects Review Workshop*, November.
9. Department of the Environment (1976) *Building Regulations*, Parts F and FF, HMSO, London.

DISCUSSION OF CHAPTER 9

R. Cullen (Cullen Carter and Hill). Much has been said recently about the development of exciting new forms in housing and yet it has been disappointing to see some of the dull, miserable, mediocre, ugly forms produced by government departments as 'experimental housing'. It is not experimental housing, it is old housing with bits tacked on. This has nothing to do with tackling the real problem of energy in the mass private-housing market. Why isn't there an injection of original design? Because professionals have little involvement in the design of speculative housing. Here is the biggest potential work load opportunity for us as integrated teams of professionals. If the same thought and application injected into hospital design could be put into designing housing real achievement would be possible.

O. S. Nielsen (Property Services Agency, currently at Building Research Establishment). The 'ugly houses' at the Building Research Establishment should not be seen as design studies; they are laboratories in which are being tested a number of solar gain devices, heat pumps, etc. In the Design Division at BRE there is currently at the design stage a design and application project where the designs introduce many new ideas such as the utilizing of solar gains at the right time and in the right location. Eventually the houses will be built.

Dr S. J. Wozniak (Building Research Establishment). Recently there has occurred both within Europe and the US too much emphasis on producing architecturally expressive designs for low-energy houses. Few of these could perhaps be seriously considered as prototypes for mass housing in the future. It should be recognized by now that intricate detailing and extravagant designs can give the most trouble in the long run. A further disappointing feature of much of the work on low-energy houses is that we have little idea whether many of these buildings fulfil their designers expectations.

R. Cullen. I believe that integrated design from absolutely first principles is the way to proceed if we are to develop a building envelope and an internal environment which operate together using ambient energy on the one hand and front end technology on the other.

Obviously, a great deal of useful and interesting analytical work is being done, but I am disappointed that during the last five years, very little prototype housing has been developed. We are still building 200 000 or so houses a year, which represents a very considerable part of the total building industry's output. This very considerable investment is being made with virtually no research for prototype work. Prototype development on an integrated design basis with professionals working together is the proper way ahead.

T. J. Wyatt (Brown Crozier & Wyatt). It has been suggested that by increasing the glass area on the hospital nucleus design (Chapter 5) from 25% to 35% a cut could be made in the lighting energy used. This is at odds with our own investigations which have shown that daylight remains a very expensive commodity. I am concerned that the impression might be given that much may be gained from passive solar radiation which comes in at low angles and yet does not cause problems of overheating, and therefore areas of glass in buildings should be increased. I am very committed to the view the reverse is the true position. Perhaps Professor O'Sullivan would like to comment.

Professor O'Sullivan (UWIST). We are talking about how we as a professional group deal with the problems that result from our endeavours to do the right things. While generally these are problems of doing some things successfully they may also be problems of failure. Within that context I believe that for the simple building, not ventilated or air-conditioned it is not easy to get the benefits from ambient gains and the problems associated with those benefits are concerned with thinking more about design rather than obtaining more measured data on quantities of solar radiation and its distribution patterns on facades.

On the hospital glass area (see Chapter 5) we are convinced from the evaluations and calculations made that increasing the glass area is a valuable thing to try within the total context.

J. Harrington-Lynn (Department of the Environment). The correct balance between lighting, solar gains and glazing requires estimation of the right balance for the particular design, taking account of daylight levels, switching patterns for the lights, the occupants expectations and the degree of summer overheating which can be tolerated. There are no firm rules for any building types. The design exercise has to be done properly in each case.

Dr J. C. McVeigh (Brighton Polytechnic). Professor O'Sullivan in presentation sketched a house plan with stove on the left and kitchen/eating area on the right. This reminded me very much of the traditional cottage in the Outer Hebrides or off the West coast of Ireland where in an investigation almost identical floor plans were found. The energy loss from such a house was discovered to be rather less than the then latest Building Regulations (1978) for a house with similar floor area. I feel it is always worth looking back at traditional solutions to the energy problem.

Professor P. O'Sullivan. If our building stock is examined it can be seen that the effect of our knowledge of the value of insulation is reflected both in the way buildings are put together and in the way investment is made in buildings in terms of expected return. Looking at ambient gains over the various building types, with certain notable exceptions, it is very difficult to find any evidence in their design that ambient energy has been taken into account or that investment has been made in a way which reflects the gains that might come from ambient energy. I find this not too surprising but rather depressing. This is an important current issue. In housing for example we must insulate houses by law but there is no law to control the energy within them. Clearly a view will have to be taken on whether this state of affairs should continue or whether the law should be extended to embrace the energy within the house and how this should be effected. I would suggest that these changes will be influenced more by the effects of ambient energy gains than by thinking on insulation. There are some important issues dependant on our view of ambient gains but the evidence that has been produced towards resolving these issues has been confused.

Dr S. J. Wozniak. The integration of architecture and engineering was given considerable thought in the ABK housing at Basildon. Published predictions* for the energy savings show that purely architectural details such as layout and orientation are unlikely to have a predominant effect.

* *Architect's Journal* (1979) October.

I agree with Professor O'Sullivan when he states that cooking loads in houses are high at the present time. However, if in the future energy became expensive then the amount used for cooking could be reduced markedly. There is no logic, therefore, in designing houses to take special account of these large gains.

Dr A. F. C. Sherratt (Thames Polytechnic). I wonder if Professor O'Sullivan would like to comment on the thermal mass of the structure. It isn't just a matter of insulation but whether an ambient gain can be used or not. Low internal mass promotes overheating and rejection of ambient gains rather than using them. For example, timber-frame housing is becoming more popular producing a very low-thermal-mass dwelling. Is this not a design point which is anti-productive to the use of ambient energy.

Professor P. O'Sullivan. It is actually very difficult to produce what is effectively a low-thermal-mass building although there are a few exceptions. Almost all buildings have a great deal of mass in them, the design decisions relate to where the mass is best situated for thermal purposes, for example, is it in furniture and fittings or is it in the fabric and floors of the building. The decisions, therefore, relate to the storage mechanisms, i.e. how and where should heat be stored. Surprisingly often these matters are neglected, the question of heat being considered as purely an internal problem. The conventional logic leads to a requirement for heat storage in the walls or their equivalent – perhaps a large water tank linked to radiators – floor storage has the disadvantage of being more difficult to get heat in and out and furniture is of little use. There is a lot to be said for this approach but forgotten in the design is that we don't just live in buildings, we also live outside. The great thing about ambient gains, particularly the passive solar gains, is that the designer can decide how much gets into the building and how much stays outside and what happens to the ambient energy remaining outside. Careful consideration to outside design using ambient energy to produce better microclimates can both improve the quality of life considerably and reduce conduction losses by reduction of inside/outside temperature differences.

Dealing with the inside of the building we know a great deal about. Relating the inside to the outside has been neglected – it does not have great importance of course if your job is purely to design the internal system.

J. Keable (HELIX Multi Professional Services). In all the low-energy houses we have been involved in we have raised the question of cooking. Usually it has not been discussed very seriously because it is the man doing most of the discussing, he is going to pay for the house, and it is he who is fascinated by energy so he wants a low-energy house. The woman of course decides in general what cooker is going to be connected and by and large the connected load of the cooker is either as big or in some cases actually greater than the connected load for the heating!

If ambient gains are to be experienced in a building (house or office) being occupied during the day when the ambient gains from solar are received it is necessary either to increase thermal mass to an exaggerated extent or completely to decouple the thermal mass of the building from the occupied space nevertheless having access to it. The latter is not a very easy thing to do but I believe it could be the way ahead in many cases.

Dr J. Twidell (University of Strathclyde). Does Professor O'Sullivan advocate a policy of double entrances and bathroom extraction ventilation as compulsory for new buildings and retrofitting?

Professor P. O'Sullivan. Although Dr Twidell's suggestion is quite a reasonable one I would not suggest a particular solution that should be pursued but that taking the evidence available and looking at inside spatial relationships of our buildings, we design afresh. One solution could be the one suggested, another might be reorganizing the spaces. There are a whole range of solutions that emerge once we actually start thinking about relating the spaces *in energy terms* as opposed to just insulating the building in energy terms, or designing the system in energy terms and believing the space relationships, whether in houses or schools are themselves sacrosanct.

10

Marketing Opportunities in the Developing Countries for Ambient Energy Systems

P. D. Dunn

INTRODUCTION

The paper discusses some possible applications of renewable energy in the developing countries. First, it is helpful to see the place of the developing countries in the world energy scene. In energy terms we can regard the world as divided into four groups:

(a) The energy rich/developed, for example, the USA.
(b) The energy poor/developed, for example, Japan.
(c) The energy rich/undeveloped, for example, Saudi Arabia.
(d) The energy poor/undeveloped, for example, Sudan.

Currently, almost all the world energy is supplied by fossil fuel (this refers to commercial energy; non-commercial energy, particularly biomass, contributes as much as 15% of total world energy use) and it is sensible that the major energy users, that is, countries under (a) and (b), should be considering alternative energy sources, particularly solar (direct and indirect) in order to substitute for fossil fuel.

The rise in oil prices has been particularly troublesome for countries in (d). This is brought out in Fig. 10.1 which refers to Thailand and which shows how imported oil prices have risen in relation to rice, a major export, over the last decade or so.

The developing countries themselves are dual societies in which there is a developed, largely urban, sector together with an under-developed rural sector in which most of the population live. The energy requirements of the developed sector are similar to that of other developed economies and it is here that most of the commercial energy is used (Fig. 10.2). In this urban area there are obvious

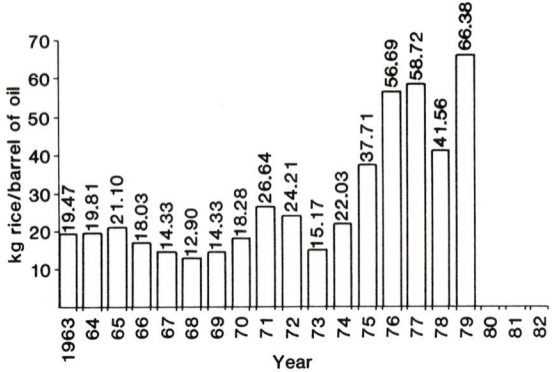

Figure 10.1. Effect of oil prices on Thai rice return (kg of rice which must be expected to buy 1 barrel of oil [1])

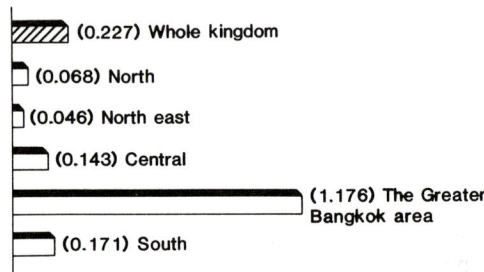

Figure 10.2. Energy consumption per capita (tonnes coal equivalent/year) in Thailand

applications for solar energy in building design and for low-temperature water heating for use in hotels and hospitals and also in refrigeration and air conditioning. These requirements are not different from those elsewhere. There is a general need for electricity and there are often considerable advantages in its local generation from alternative energy sources. In the less-developed rural areas energy needs tend to be for dispersed supplies of 10 kW or even less. The problem of rural energy supplies in developing countries has been discussed elsewhere [2] and will not be referred to further in this paper though it is of great importance. In this chapter we will look at small-scale electricity supplies of the order of 100 kW.

SMALL-SCALE ELECTRICITY SUPPLY – GENERAL CONSIDERATIONS

There are many communities which at present rely on diesel-electric generation. One problem often experienced is that of unreliability in fuel supply, a second difficulty is the increasing cost of the fuel. In looking at alternatives it is necessary to consider a number of factors which include the nature of the load, the financial resources of the consumer, the technical possibility of alternative energy sources, the selection of sites, the requirements for storage, the technical skills necessary for maintenance and repair. The conventional alternative is usually the extension of grid connection.

WIND-GENERATOR, DIESEL, BATTERY SYSTEM

One method of reducing fuel cost is the provision of a mixed system in which diesel-power is supplemented by a wind generator. Load balancing is provided by lead–acid battery storage. Reference 3 describes such a study which has been carried out for the Scottish Island community. In this system the peak power is 100 kW and the load variation is given in Fig. 10.3. The study includes the

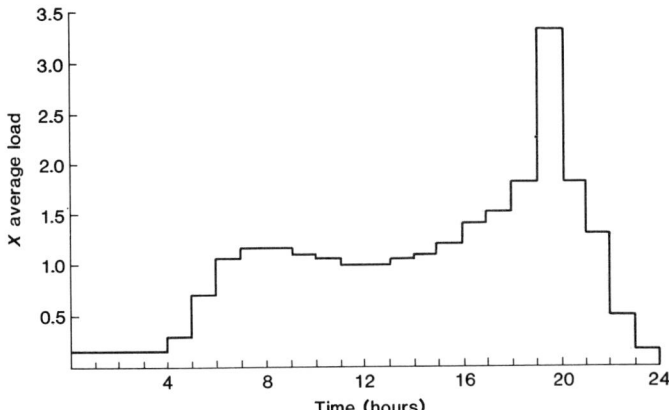

Figure 10.3. Daily load profile

optimization of the windmill/battery/engine size. Figures 10.4 and 10.5 indicate how the cost varies with windmill swept area. The preliminary results of this study indicate that generation cost can be reduced by a factor of 2 by the introduction of wind power. It is important to emphasize that these results are illustrative only and should not be taken out of context. Nevertheless, they are sufficiently promising to justify serious consideration of such a system where the wind regime is favourable.

SOLAR POND

This development may well prove to be the most important application of solar energy in the tropical countries. A solar pond is a shallow pond, typically 1–2 m

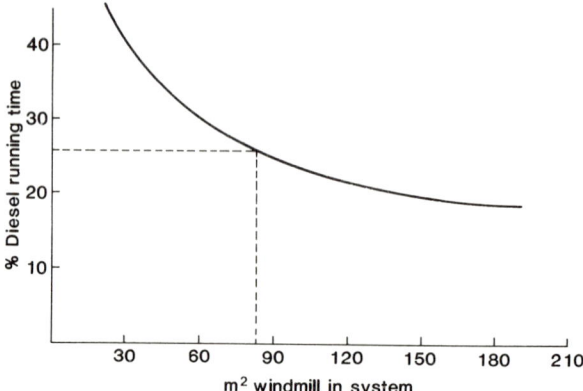

Figure 10.4. Diesel running time versus windmill size for 10 hours storage

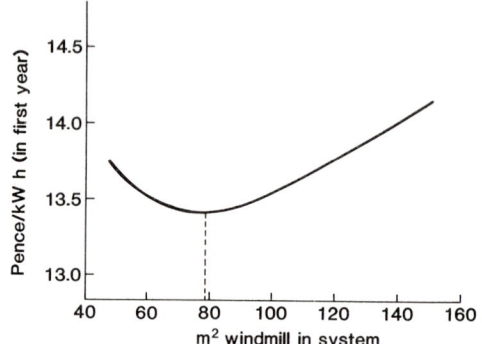

Figure 10.5   Electricity price versus windmill size (storage = 10 hours of mean load)

in depth and having a blackened bottom (Fig. 10.6). The long-wave solar radiation is strongly absorbed near the surface but around one-quarter of the total incident radiation is absorbed at the bottom of the pond. The convection which would normally occur due to heating of the bottom layer is avoided by arranging for a strong density gradient from bottom to top of the pond by the use of dissolved salt. The bottom of the pond is a saturated solution and the concentration decreases progressively to the top. Convection is suppressed and the temperature difference between top and bottom is typically 25–30°C at the top and 80–90°C at the bottom. The hot layer is caused to flow through a heat exchanger and the resulting thermal energy operates a Rankine cycle engine which in turn drives an electric generator [4].

Much of the work on solar ponds has been carried out in Israel where a 150 kW(e) installation has been reported in operation. Because of the high heat capacity of the pond it can operate for several days or even weeks without heat input.

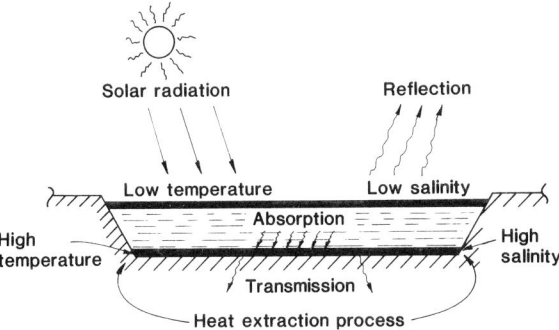

Figure 10.6. Solar pond for power generation

HYDROELECTRIC SYSTEM

There is considerable potential for the introduction of small-scale or micro-hydroelectric generators in many areas of the developing world. Many suitable sites exist in South America, Africa and South East Asia. The maximum power level required will range from kW to MW. The best sites for such small-scale generators occur in mountainous or hilly regions on the upper reaches of rivers. They are not the sites normally considered in assessing the hydroelectric potential of an area. Such surveys are generally restricted to the selection of sites suitable for large-scale electrical generation and occur only on major rivers. The largest national programme on small-scale hydro-generators reported is in the People's Republic of China [5, 6]. The programme was started in the 1950s and by 1972 some 60 000 stations had been installed with a total capacity of 3000 MW and an average output of 40 kW per station.

The unit cost of electrical energy will decrease as need and power level are increased. Cost is also dependent on the degree of control required over the output power level (and frequency if a.c. power is generated). The turbine design selected will depend on the head of water. For high heads (20–200 m) Pelton runners will be used, these are relatively simple to manufacture. For intermediate heads (5–50 m) a radial Francis turbine is selected; this is somewhat complex to manufacture. At low heads (2–20 m) propeller turbines are used, the conventional large-power turbine is the Kaplan but cheaper designs are available for lower power levels. Generally effective and reliable commercial equipment is available in the power range required.

The selection of the site and its evaluation is an important aspect in the planning of a hydroelectric system and can represent a significant fraction or even the major component of the total cost. Nearby rainfall, river flow and level data are required together with estimates of the minimum and maximum flow over several years. Typical projects are described in Reference 7.

OPPORTUNITIES AND BARRIERS TO IMPLEMENTATION

There are commercial opportunities in the small electrical-power generation field in the areas described in

(a) consultancy
(b) generation plant supply, installation and commissioning
(c) manufacture plant supply, installation and commissioning in order to enable some or all of the generation plant to be manufactured in the user country.

In order to have credibility in these areas it is necessary to have

(a) trained personnel
(b) manufacture experience
(c) field experience.

There are some experience and personnel currently available in the UK and with the development of new energy programmes such as in wind power, these will increase.

SUMMARY

There are several developments in renewable energy which offer the prospect of economic electricity generation either in combination with the conventional diesel generator or in place of it. As fossil fuel prices rise these alternatives will become increasingly competitive. Two of the systems described, solar pond and hydro-electric, are independent of an external supply of fuel. In all cases there is a significant local input to the total system in site preparation and also the possibility of the local manufacture of some of the plant. All require considerable expertise in the site selection and design together with provision of special equipment such as vapour turbine in the solar pond. They all three offer the possibility for our industry to make a worthwhile contribution to the development of these areas and have the advantage of involving significant participation.

REFERENCES

1. Poonsab, S., personal communication.
2. Dunn, P. D. (1978) *Appropriate Technology*, Macmillan, London.
3. Sexon, B., Slack, G., Dunn, P. D., Lipman, N. and Musgrove, P. J. (1980) *Wind Turbine Diesel Electricity Generation for Small Remote Communities*, Energy Group Report, University of Reading, UK.
4. Tabor, H. (1980) *Solar Power Generation – The case for Solar Ponds*, Report for The Hebrew University, Israel.

5. Dickenson, H. (1972) *Rural China. Occasional Report*, School of Engineering Science, University of Edinburgh, UK.
6. Dickenson, H. and Whittington, H. W. (1978) *Rural Electrical Supplies*, School of Engineering Science, University of Edinburgh, UK.
7. Lawrence, E. D. and Pilgrim, J. W. (1979) The transfer of small scale hydroelectric technology. In *Conference on Small Scale Energy for Developing Countries*, Reading University, UK, 8–9 January.

DISCUSSION OF CHAPTER 10

Dr S. J. Wozniak (Building Research Establishment). There are other resources apart from energy. It might be a mistake to concentrate exclusively on the conservation of known reserves of one resource. For example, world reserves of high-grade copper ore are finite. Processing of lower-grade ores entails greater energy expenditure and in many cases greater environmental impact. It is a mistake to consider that energy is the only resource worth conserving.

Dr Dunstan stressed in Chapter 1 that there was considerable scope for wasting resources within the building industry. In particular, in applying new technology, it could be better to make ten mistakes and investigate them instead of having to repair 10 000 faulty systems. Dr McVeigh (Discussion Chapter 11) advocates building more new designs of houses each with many novel energy conservation features. This is certainly a very exciting idea and I agree that in some cases simple monitoring can be useful. However, the real problems may be a little more difficult and may require a more organized approach. It would be helpful if we could know the cost-effectiveness of each design change and of each new component. Future optimization will only be possible if we know the long-term average useful energy contribution of each component within any particular system. It will, therefore, be necessary to monitor new designs particularly well, especially if they incorporate more than one novel feature.

Professor P. Dunn (University of Reading). Dr Wozniak's comments would also apply to electric cars and lead. It should also be borne in mind that each development has itself an 'energy' cost in the energy required to bring it into being. For example, the solar satellite, proposed by the Arthur D. Little Company where 8000 MW of electricity would be generated and beamed down would require so much energy that optimistically it would take over six months operation to get the energy back – if the estimates are wrong, much longer. A great deal of care has to be taken, particularly so with silicon cells which in the present form are very energy intensive.

Dr J. Twidell (University of Strathclyde). A bad example is the way we make cars that rust away in about seven years. It is monstrous that *our* money (and I would emphasize government money is *our* money) is being used to subsidize a major manufacturer to make cars that collapse after about seven years. I would support a subsidy to make cars capable of lasting twenty years (as they used to) and also capable of 100 miles per gallon. Such a policy is part of a resource overview or ethic. Because we have no resource strategy we have resource stupidity.

Dr S. J. Wozniak. The corrosion rate experienced in most modern motor vehicles can be dependent largely upon maintenance. This illustrates one of the essential difficulties for conservation of resources. In the developed countries the cost of labour is very high compared with the cost of energy and other raw materials. This is a crucial factor and one that has perhaps not been recognized sufficiently. Nevertheless, we appear to have recognized that high comfort standards are not immutable and that I find encouraging.

D. C. Pritchard (Polytechnic of the South Bank). My son is a motor-vehicle engineer in the South Sudan. Because the ordinary lead–acid accumulators lose their charge overnight in the Sudan a man-made hill has been made so that by leaving a vehicle on top at least one vehicle can be started in the morning. Is this compatible with the idea mentioned by Professor Dunn of using windmill generators to charge lead–acid accumulators?

Professor P. Dunn. A splendid solution, precisely the right solution to the problem. We are working with the Sudanese on a major small windmill programme, which initially will be for pumping but it is intended to extend it to electricity generation and lead–acid batteries will form part of the scheme because other battery types are too expensive. Lead–acid batteries do lose their charge but the problem referred to may be a question of adequate maintenance.

11

Ambient Energy – The Resource of the Future

John Twidell

INTRODUCTION

The brief given for this chapter included a request to provide 'a forward look at the likely application of ambient energy in all its forms in the future (30–100 years on); the paper to be a personal appraisal of the way technology and social variables are likely to change'. This was a welcome invitation, and one tackled with enthusiasm from a personal background in the Third World and in Scotland. I indeed accept that ambient (that is, renewable) energy has to be *the* resource of the future, but this can only occur with a radical change in the standpoint of established energy authorities. I emphasize the international component of our thinking. As the Brandt Commission has stressed [1], our world is a unity – a complex yet integrated system of ecology, economies and, I hope, justice. Justice is here used to encompass fairness in the distribution of resources, legislation to ensure the maintenance of ethical and technical standards, and human compassion in the process of development and innovation.

On a recent solar energy project in East Africa, our research group met a wide range of opinions regarding future energy supplies for countries with rapidly growing populations now heavily dependent on oil. Many opinions were extremely pessimistic, and it certainly did not help to join this pessimism with comments on the outlook facing the developed world. It is not difficult to be pessimistic regarding future energy supplies, but I entirely reject this doomsday approach – not because it is necessarily wrong, but because the energy future can be optimistic so long as mankind behaves in a moderate, caring and scientific manner. I hope to convince you that there will be sufficient high-quality supplies available for all mankind, but only if we develop renewable supplies with agreed ethical, scientific and economic foundations. It will not be good enough to provide a shoppers' guide to general energy options as if one purchase is roughly equivalent to another. The role of energy supplies is far more influential than that. One purchase is usually not the same as another – they are not alternatives. Renewable energy is not to be developed at the toss of a coin; there are basic principles at stake of far greater significance than superficial choice. Any meaningful view of the future must take these into account.

ENERGY USE

Energy is vital because all life and technical processes require a flow of energy and a redistribution of materials. The biosphere exists using the steady flow of solar energy that maintains the Earth's temperature and acts through photochemical processes to maintain life forms. For thousands of years man existed using the almost immediate forms of this ambient (renewable) energy, and then about 200 years ago industrial man discovered intense sources by releasing stored energy from fossil fuels. This bonus became the driving function for the development of the industrial revolution that has allowed the coal-producing countries to dominate as the developed nations. Note that the few countries that had coal, still have it, and will continue with assured supplies for several hundred years. These supplies, although essential for the coal-producing countries themselves, cannot meet world energy needs either in amount or at a satisfactory rate.

The main requirements for energy (apart from food) have always been heating for cooking, warmth and industrial processes; transport; and machine use. Electricity has become an energy vector of extreme importance because of its 'high quality' and control in operating machines and devices. In general the energy fluxes associated with national electricity supplies are small however, and the importance of electricity is more associated with the availability of electrical potential than high-power usage. We must not equate 'electricity' with 'energy', nor must we welcome the stupidity of turning thermal energy partly into electricity only to be directly dissipated for heat. The best use of electricity is undoubtedly the transmission of energy from immediate mechanical sources such as wind, wave, tidal and hydro power. Electricity is important, but there is far more to energy than electrical engineering.

More recently, and somewhat transiently, petroleum fuels have become available. These have provided a further driving function for a second spurt of development in the older industrial countries of the 'North', and for a primary burst in the remaining countries of the Third World developing 'South'. Note, however, that nowhere in the world have the long-used renewable forms of energy lost their value, so that today the total use of renewable energy is about one-third of total world fuel use. This is principally because the majority of people cook with firewood or other biomass products, and because hydro power is universally prized. If we were to add to these renewable flows the gains from ambient energy use in buildings of temperate and colder regions, then the total reliance on renewable energy now is of major importance. Please dismiss any idea that ambient/renewable energy is a new idea without present day experience.

The world has, however, reached a crisis with fossil fuels, especially with oil. The reasons for this can be explained with a single equation:

$$R = S \times N$$

Here R is world commercial energy use, S is the average per capita commercial use, and N is the user population. The crisis occurs because of three concurrent factors. First the number of people (N) expecting to use commercial energy is doubling about every twenty years due to absolute population increase combined with an increasing proportion educated and constrained to use commercial energy. Population control and educational stabilization may eventually stop this growth rate, but N will not become constant in the foreseeable future unless severe global distress is the cause. Secondly, individual expectation of improved standard of living is attempting to force per capita use S to also increase exponentially. The resulting pressure on world commercial supplies R can only be satisfied by exponential growth of $\sim 5\%$ per annum, which is a doubling time of ~ 15 years. Thus it would take all the running world energy growth could do, just to keep average per capita supplies constant at the abysmally low present levels. If this were not bad enough, the third concurrent factor now appears – the depletion of the finite resources of petroleum fuel. This causes price rises and disruptions that remove the developing potential of oil even further from the world's poor. It is only on a world scene that the full impact of the energy crisis can be seen, and it is to the credit of the Brandt Report that the implications have now been put starkly before world leaders, so that no one is without excuse for not realizing what is happening.

Nuclear electricity

Some people see a world energy solution in terms of nuclear energy. However, if ambient energy is to be *the* resource for the future, then clearly nuclear energy cannot be also. Nuclear energy (incorrectly called atomic energy) can realistically only supply electricity. Electricity forms only a small part ($\lesssim 10\%$) of national energy needs since transport and heat supplies dominate. Most countries do not have grid electricity networks or the electrical infrastructure to make use of intense centralized sources of electricity. Indeed increased centralization is contrary to most development plans. The only long-term nuclear policy is based on breeder reactors, and both these and conventional reactors are fraught with difficulties of security, weapons proliferation, safety, short lifetime, costs and waste disposal. In any case, the developed countries tend to have an abundance of electrical generation capacity and future needs can be met by greater efficiency of use rather than increased generation (e.g. Scotland has arguably 80–100% overcapacity if one allows for sensible usage – data are available in the annual reports of the SSEB and NSHEB). Career opportunities after construction are for the highly skilled with little general employment. The developing countries have great need for electrical supplies, mainly for low-intensity distributed use for lighting and machines and not for heat. The very high development costs of complete nuclear programmes are preventing adequate support for the development of renewable supplies. Perhaps the greatest mistake of all is that nuclear energy is a physical science answer to a multidisciplinary ecological problem. Energy is not the sole preserve of the physicist and electrical engineer, and supplies are needed to match all the needs of the community and not just electricity.

Solar energy

Why am I so confident of the ultimate success of renewable energy? It is because of two essential physical properties of solar radiation linked to the first and second laws of thermodynamics.

(a) The total solar flux onto the Earth averages 30 MW per person of continuous power. Used efficiently a satisfactory per capita rate of energy use is about 3 kW per person, and so we need only harness about 0.01% of the solar flux to have energy sustainability.

(b) The solar flux arrives with the highest thermodynamic quality from a temperature source of nearly 6000°C. The flux is a photon flux at visible and near-visible frequencies. Solar photon-radiation is, therefore, an excellent source for high-quality energy generation (e.g. chemical products and electricity) and for work. The high-temperature source of the Sun can be utilized directly in photo-physical processes to produce work. The efficiency of direct photon-induced generation is potentially far higher than that of solar thermal machines where the radiation is degraded to heat in the collector. Note too the importance of photo-chemical products for material use and for energy storage. Photosynthesis is the most neglected of processes and we need to rapidly develop photo-chemical application to harness the full potential of solar energy. Scientifically, present day commercial energy generation is as chalk compared with cheese in comparison with solar photon generation of energy and chemical products. Oppenheimer is said to have admired Teller's scientific development of the fusion bomb as 'a sweet solution'. For me, nothing can be so sweet as the scientific principles of photosynthesis.

Support for solar energy is, therefore, support for science fully compatible with life processes. The crudities of present day intensive engineering result from attempts to isolate mankind from the science of our natural environment. Such policies are dangerous, and may lead to a total break with life.

Summary

We have a crisis in commercial and non-commercial energy supplies at a world development level. I am convinced that fossil fuel and nuclear sources are entirely inadequate for meeting global energy need both now and in the future. Fortunately solar energy is more than adequate as an energy flux and as a source of high energy potential. As yet this is mostly a theoretical potential – how can we develop the resource to its full value?

RESOURCE ETHICS

Introduction

Ethics is as fundamental to energy supply as it is to all mankind's endeavours. Whatever the supply, we shall either be draining finite resources or tampering with our environment, and such action should not be undertaken without thought for wider implications. It is possible to advance piecemeal with renewable energy development viewed merely as a fuel saver, or indeed as an insurance against national coal strikes and nuclear accidents, but such policies are insipid and stultifying. This is probably the UK's attitude at the moment, and in quantitative terms is represented by the mere pittance of 4% of energy R&D funds. Significant progress will only be made towards a full development of ambient energy if those of like mind join forces and work with a common aspiration. First we must accept a common set of ethical values, then must follow good science, and finally we must decide between options by economic criteria. Please note the order of development – ethics, science and lastly economics – certainly not the reverse.

Resource ethics outlined

The ethics we must base our commitment on, I call 'Resource Ethics'. The subject does not as yet exist as an established body of thought, yet the anti-materialistic aspects are probably in common with all established ethical and religious philosophies. The importance of resources has been apparent over the last twenty years in association with warnings about mankind's harmful impact on the natural environment, with quantitative assessments of the depletion rates of finite material and energy supplies, and with a growing awareness of the uneven global distribution of raw materials and social benefits. In the UK, recognition of such attitudes is apparent, for instance in the formation of the Department of the Environment, the Department of Energy and the Nature Conservancy Council. In education, from university to primary level, quite fundamental changes have occurred in response to resource and environmental issues. All this is good, but the total effect is insignificant in comparison with the full demands of Resource

Ethics. Let us summarize some principles that must be accepted before a full commitment to renewable energy can occur.

(a) The Earth has resources of energy and materials that must be managed for common use by all mankind. Some of the resources are renewable and therefore continuously available, other resources are finite. Energy and material resources are linked, and either may become limiting on the processes of life and development.

(b) Man is part of the biosphere, and we must aim to live in harmony with natural ecology, and not at variance or in isolation from natural processes.

(c) To the recognized individual human rights of food, shelter, health, education and freedom, must be added the right of participating in a meaningful individual share of the Earth's resources.

(d) Finite resources, if used, will eventually become unavailable; either because of depletion or because of high price. Therefore it is our duty to use finite resources efficiently, moderately and cooperatively. This is the only responsible attitude to fellow mankind, both for now and in the future.

(e) Such principles must be embodied in law. In addition, all those engaged in harnessing the Earth's resources should act according to an equivalent of the Hippocratic oath. This 'oath' should state the practitioners' duty to safeguard resources in a manner above pecuniary and materialistic self-interest.

Application

Consider some examples of action that would follow from accepting such principles of Resource Ethics.

(a) It would be fundamentally wrong (and illegal) to generate electricity from a finite resource and allow two-thirds of the power to be immediately wasted, when other options exist for its use. For instance with generation from coal, we should not so despise the value of organic material, the high thermodynamic quality of electricity and the labour of miners, that only 30% is usefully used at source and only about 15% is usefully used by consumers. The Marshall report [2] on combined heat and power gives support for this attitude. Local district heating, horticulture in heated glasshouses, fish farming and the supply of process heat directly in industry are all realistic options for increasing the efficiency of use of thermal electrical generation. Therefore, planning resource use from an ethical standpoint need not lead to scientific or economic stupidity. It does, however, certainly save time.

(b) Accepting an ethical view of resource use immediately leads to the development of efficient energy use and conservation.

(c) Concern for the depletion of finite resources immediately leads to consideration of renewable sources and a more enthusiastic thrust for their development.

(d) With housing it would be fundamentally unacceptable to allow building that did not meet optimum standards of energy-use efficiency and that did not harness local renewable supplies (e.g. sunshine). It would be a legal requirement to include such criteria in the planning process.

Some of you will accept this philosophy of Resource Ethics, others will not. To those that do not I offer a challenge – show me a long-term strategy for world survival and sustainability that is not forced to accept the aims of these principles.

SCIENCE

Renewable energy

Independent of and secondary to the ethics of resources, is the science of resources. In the case of renewable energy there are a number of distinctive scientific principles that must be appreciated if increasing use is to be made of this resource. Without this appreciation there can be little progress. At the start we need a definition: 'Renewable energy supplies are those supplies obtained by tapping into the various flows of energy occurring naturally and steadily in the environment'. This definition contrasts with that for non-renewable supplies, which are obtained by initiating the flow from some form of concentrated energy store. Immediate contrasts result between the two energy supplies.

(a) Renewable input fluxes are low ($\sim 1 \text{ kW/m}^2$ maximum) and dispersed. Centralization and concentration of the fluxes by engineering systems will be difficult and expensive. Non-renewable input fluxes are usually intense ($\gg 1 \text{ kW/m}^2$) and are concentrated at source. Dispersal and distribution by engineering systems may be difficult and is expensive. Thus renewable energy supplies form a bias to a dispersed form of society (~ 2 people per hectare as a very approximate guide).
(b) Each form of renewable energy flow has distinctive scientific and dynamic aspects. Specialist and interdisciplinary study is necessary to develop useful energy supplies. The range of these aspects is extremely large and far greater than required for non-renewable supplies.
(c) Renewable energy generation will be dependent on the properties of the local environment and climate. There will be no single common system of use applicable throughout the world or even throughout one region. There will be great diversity leading to varied opportunity and development of robust systems. Some aspects of particular devices will be in common between locations, but always there will be the requirement to adapt to local conditions and supply needs. For instance wind generators developed for Northern Scotland would not have to be so robust in most other parts of the world, and would be designed to match lower wind speeds. Non-renewable supplies are generally independent of location. Mass production of common devices for global use is favoured. This has many advantages, but can lead to

fragile systems (e.g. national strikes, fault avalanches, terrorist activity, price monopolies).

(d) Renewable energy systems have zero fuel cost, reasonable maintenance requirements and high capital cost compared with fossil-fuel systems. Non-renewable fossil-fuel systems have fuel costs increasing in an unknown manner, easy maintenance requirements and relatively low capital cost. Non-renewable nuclear systems have low (but unpredictable) fuel costs, as yet unknown reprocessing and waste requirements, and high capital cost.

(e) Renewable energy systems have predictable danger; all are essentially safe when broken. They have very low and usually negligible effect on the natural environment. Aesthetic criteria dominate when considering their environmental impact. Non-renewable systems often have unpredictable danger, and are frequently more dangerous when broken. There can be extreme damage and perturbation in the natural environment by effluent and waste. These pollution-related aspects dominate when considering environmental impact.

(f) The social and employment aspects of renewable energy supplies are in general benign and akin to agriculture. There would seem to be a bias to employment opportunities of a wide variety. The social and employment aspects of non-renewable supplies are generally known. Increasing centralization and complexity would seem to bias the system towards very specialized employment opportunities.

(g) Renewable energy systems tend to produce self-reliant and independent communities with a bias to small-scale application. For the majority of the people of the world, who have always received essential supplies of energy from renewable resources, development of renewable supplies will lead to improvements in life style and quality. Non-renewable supplies tend to produce centralized, structured societies. For the majority of the people of the world this leads to considerable social and cultural disruption. Agricultural communities are particularly vulnerable to disruption by non-renewable supplies, both in their inception and depletion.

(h) We must consider the *need* for energy before the supply. In the UK predominant energy needs are for space and process low temperature ($\sim 80°C$) heating ($\sim 39\%$); high temperature heat (28%); for transport (23%), and only 10% needs to be controlled 240 V 50 Hz electricity [5]. The best efficiency results from carefully matching supply to need. In general, poor efficiency results from producing high-grade energy (e.g. electricity) and using it only for low-grade use (e.g. space heating). There are other scientific principles that apply to energy systems.

(i) Energy flows from a source (generation) to a sink (need). The only meaningful efficiency is for the whole system. At present, the efficiencies of most of our systems are abysmally low. Consider electricity use efficiencies: generation (30%); transmission (90%); and use ($\sim 30\%$); total efficiency 8%.

Consider house design [3], the average non-ambient heating load for new houses built according to building regulation standards is 46 GJ per year, and that for a 'solar structured' house with passive heating is expected to be 16 GJ/year. Thus present houses waste 30 GJ/year and have an equivalent efficiency of 35%.

(j) The control aspects of present-day energy systems, especially at end use, are crude and largely untouched by modern microelectronics. The opportunities for improvements are legion, especially for electrical engineers.

Application

Consider now application of the scientific principles we have established. What kind of planning can we produce associated with ambient/renewable energy?

(a) Planning must be regionally and locally based, and will be highly dependent on local environmental factors; local needs and local social factors. It is therefore a fundamental mistake to have a single national office charged with renewable energy development. It makes even less sense to have this based in the immediate environment and influence of nuclear energy authorities as is done in Britain at the UKAEA, in Europe at the Ispra Establishment, and in Denmark at the Risø Laboratories. The renewable energy opportunities in Scotland are in general quite different from those in Southern England. Clearly the place to investigate, develop and encourage Scottish renewable resources is Scotland. Likewise, since the agricultural and rural influence on renewable energy systems will be large, central London is probably the worst possible place for developing renewable energy supplies.

(b) The optimum manner of developing renewable energy systems for generation and use is at small scale in remote and rural locations. In this way the most efficient matching of supply and use can be made. It is pointless to consider scaling factors of generation only, without the scaling factors of use. Thus whole system efficiencies can be surprisingly large when considering local use from small-scale local sources. The UK policy of supporting very large-scale complexes for electrical generation to 10 000 MW capacity is going against the sensible use of renewable energy in the dispersed environment and the principle of matching supply to need.

(c) The richness of opportunity exploiting renewable supplies is immense. It is scientifically inadvisable to make national decisions to favour only certain energy options and exclude others. These decisions must be taken at the local level. Consider hydrogen for instance. Hydrogen is an excellent chemical energy store, capable of production from a number of renewable sources and very versatile in use. It can be generated locally and used locally (e.g. a 10 kW household aerogenerator can produce hydrogen to store excess power, and the fuel used for cooking, heating and transport). Yet the UK national decision not to favour research and development in hydrogen is used as a reason for not favouring hydrogen for small-scale local use. Here is an example of centralized decision making being unable to cope with non-centralized and varied opportunities.

ECONOMICS

Introduction

Never has a word lost its meaning more than 'economics'. The Greek meaning is that of stewardship and management of a household or estate. Today the high motivation for stewardship has been lost in the materialistic greed for profit making. Everything has to be justified in terms of the transient and shallow values of monetary gain. In considering the future of ambient energy I have considered monetary factors in third place after ethical and scientific factors. It is essential to keep this order, so that monetary economics becomes the servant of resource ethics and resource science, and not the master. Ambient energy is related to natural processes in the biosphere that maintain long-term sustainability. True economics should also be concerned with long-term sustainability of resources, but unfortunately today it is not. There is a place for economics, but only in selecting between the choices established by Resource Ethics and science.

An example, housing

Consider the example of housebuilding. Resource Ethics would state that: (a) maximum use must be made of ambient energy; (b) the structure should be permanent and therefore of good design and workmanship; and (c) minimum use must be made of finite resources during the (long) lifetime of the building. Following this strategy, science considers the environmental and resource implications of various designs without considering monetary aspects. For the UK as a whole, passive solar input is sure to be a dominant opportunity, and this becomes especially so in Scotland because of the heating season extending into the summer [4]. Design criteria for thermal capacity, insulation, controlled ventilation etc., all follow from the decision to utilize passive solar energy. Other possible inputs will be direct solar and perhaps wind, hydro, and biomass energy. Food production inside or beside the house must be considered, together with the proper ecological use of waste effluent. Certainly every house must have some provision for a conservatory or a greenhouse. Fish culture should also be considered. The complete scientific assessment of the possible design will allow many variations, but all must be suitable for the particular site, local environment, and expected inhabitants. Science, therefore, lays out the choices as directed by resource implications.

At this third stage, and not before, economics and local planning factors are considered. These must include present and expected factors within the lifetime of the building (at least 50 years and perhaps 150 years). Note that scientific factors will have undoubtedly already determined that houses face south, make full use of glass, have certain related structural properties, and perhaps have certain active solar devices and control systems. The exact criteria must be established within the broad framework by building regulations. Economic factors will probably determine the size of the house, the aesthetic qualities, and the choices concerning food production.

A FUTURE SCENARIO

In 150 years time, let us imagine an optimistic world scene on the assumption that a knowledgeable and committed development of renewable energy started from, say, 1990. Two factors will be outstanding.

(a) In the presently developed world, the efficient use of energy and continued supplies of coal with its subsidiary fuel products will have provided the opportunity for a smooth transition to almost total dependence on renewable energy. The effects on the restructuring of society, especially the dispersion of cities and strengthening of rural sectors are most pronounced.

(b) In the presently developing world, the days of fossil oil supplies are seen as a small historic episode in the history of greatly transformed countries. In the tropics especially, every variety of renewable energy supply has its place according to the environmental possibilities and newly developed skills of the people. Urbanization never occurred to the extent of the northern countries at the close of the twentieth century, although there is intensive small-scale, agricultural-based development over wide areas. Of particular importance is the role of biomass energy and material sources, and of household photoelectric sources.

How did this occur? Consider two countries in completely different environments and circumstances, Kenya and Scotland.

Kenya 2150 AD

In 1985 Kenya was the first country to vigorously enforce the control of tree growth and wood use, combined with large-scale deployment of efficient biomass cooking stoves. Geothermal energy and widespread use of medium- and small-scale hydro units began a dispersal of electrical supplies. At the coast, wind generators, developed off East Anglia in England and in Scotland, have proved the most efficient and long lasting of any in the world. Of particular importance has been the success of transport fuel for road vehicles based on; (a) liquid fuels from fuel crops, (b) vehicles for local use powered by hydrogen from a variety of sources including photoelectricity, wind and small hydro, (c) commitment to public service vehicles in place of private transport, and (d) development of self-reliant townships requiring little transport of goods. Per capita energy use has never been large, and the quite extraordinary developments in microelectronics have enabled communications, public entertainment and education to be satisfactorily undertaken for the dispersed populations. Metropolises of population greater than 1-million people are unknown, and so too are low-populated, individual farmsteads. The mass of the population of 60-million people live in townships of 10 000–100 000 people spread over highly developed rural areas of intensive horticultural and agricultural use. Household electricity supplies for lighting, electronics and small machines are based on 4 m² photoelectric panels orientated to follow the Sun.

Scotland 2150 AD

Of all the countries in Europe, Scotland was the most fortunate in the 1980s in having large underpopulated areas and islands obviously requiring renewable energy supplies. The extraordinary energy development of Scotland is now recorded in history – the world's most vigorous wind-energy sites, the world's optimum wave-energy potential, the world's optimum tidal-current generation, the UK's optimum hydro- and forest-products areas, many areas of relatively small-scale, tidal-range, power application due to local geographical features, the coldest summers in Europe requiring economically favourable passive solar heating, and the world's greatest expertise in alcohol fermentation. Also in the 1980s, the North of Scotland Hydro-Electricity Board took advantage of its unique charter and its experience of hydro, wave and wind power to become the world's first energy authority to power highly developed self-reliant communities. In AD 2010, the Orkney Islands were supplied with all forms of energy by the Board (including transport and hydrogen gas) generated from dispersed sources of wind, tidal power and some biomass applications. Noteworthy has been the growth of battery- and hydrogen-powered vehicles on the islands. The mechanical and electrical industries of Scotland, which were moribund in the early 1980s, have been reinvigorated by the stimulus of renewable energy options, which now even include the manufacture of large Ocean Thermal Energy Conversion floating structures for use in the tropics. The typical Scottish house now has little resemblance to those of the 1970s, the major transformation being the optimum use of passive solar gains, the sophisticated control of ventilation, and the high standards of thermal insulation. Many houses, especially the farms, make extensive use of wind and hydro power. Of the greatest social importance, has been the repopulation of the Highland and Island areas due to the development and use of local renewable energy options.

CONCLUSION

In looking to the future of ambient energy, I can see no hope for mankind on a world scale unless there is a total commitment to environmental energy use. We should not be talking about alternatives, but instead about new directions. With sensible stewardship and maximum efficiency of use of these resources, then there is adequate energy, and therefore time, to make the change of direction. I estimate this time at between 50 and 150 years, but only if good sense prevails. In this period the necessary social and technical changes can be made to put mankind at least at an acceptable level of social well being and also on the path of sustainability.

REFERENCES

1. Brandt Report of the Independent Commission on International Development Issues (1980) *North–South: a programme for survival*, Pan Books, London.
2. Marshall Report to the Secretary of State for Energy (1979) *Combined Heat and Electrical Power Generation in the United Kingdom*, Energy Paper 35, HMSO, London.
3. Bartholomew, D. (1980) Passive solar heating – the prospects for UK housing. *Renewable Energy News*, Issue 4, Department of Energy, London.
4. MacGregor, A. W. K. (1980) Solar heat for the Highlands and Islands of Scotland. In *Energy for Rural and Island Communities*, J. Twidell, Pergamon Press.
5. Leach, G. Lewis, C. Romig, S., Van Buren, A. and Foley, G. *A Low Energy Strategy for the United Kingdom*, Science Reviews and International Institute for Environment and Development, London.

DISCUSSION OF CHAPTER 11

R. Cullen (Cullen Carter and Hill). Although I am absolutely with Dr Twidell in spirit, he benefits from the protection of an academic environment. In practice for some thirty years mainly related to housing I have learnt to face the realities of life. I know we are not investing sufficient money to provide the right quality of housing. I have seen successive governments produce stupid programmes based on dubious criteria which have created the housing problem.

As professionals we should demonstrate that investment is needed if problems are to be solved on a long-term basis. There is general acceptance that we can only produce 200 000 houses a year rather than half a million a year. Similarly that schools will last longer because educational methods will not have changed so much that the schools must be renewed every fifteen years. We are thinking and talking long term and must therefore invest appropriately for that long term. We should demonstrate that by spending more to produce better quality and avoid constant expenditure and maintenance will save resources.

I would like to see the government support some real research/development projects on a wide scale over the country, investigating and demonstrating the benefits of spending 5% or 10% more than current cost limits.

P. J. Jonas (Department of Energy). Government are considering a whole series of demonstration projects for energy conservation in buildings. For example, a scheme for application of microprocessor control to commercial buildings.

Detailed discussions are currently taking place between Building Research Establishment, the Housing Development Directorate and the Energy Technology Support Unit on mounting a series of programmes in domestic housing. The results of the better-insulated housing programme where housing on six or eight sites has been the subject of a controlled investigation will be the starting point. Because government is not prepared to give large-scale grants it is very difficult to find a practical demonstration project in the housing sector. It is intended to increase activity but the cost of monitoring is enormous and unless the energy conservation or alternative energy projects in the housing sector are carefully monitored evidence on which to base further work will not be forthcoming.

Dr J. Twidell (University of Strathclyde). With household energy use, dramatic improvements can occur for virtually no extra cost if done at the planning stage. Such improvements should be compulsory, and relate to the 50–100-year expected lifetime. We are not allowed to put up houses that would fall down in this period so many regulations have to be obeyed for the sake of subsequent inhabitants, although it is not their money that pays initially for the house. The same principle should apply to energy use, but very few energy-related regulations exist for the benefit of future inhabitants of the house. Our own house is an example. It is on a south-facing slope, but was built completely back to front with the living rooms facing north. I have spent ten years trying to turn the house round, and I can assure you this is difficult and expensive! At the planning stage, a 180° reorientation would have cost nothing extra. I am sure planning authorities do not even consider the compass directions on plans so it is not surprising such bad arrangements occur.

P. J. Jonas. The government solar energy programme is to publish guidelines for passive solar heating – recommendations from a major research project are now in their final stages. The British Standards Institution is drawing up a Building Code for Energy in Buildings called a Head Code, and it is also hoping to develop a code on energy in housing, complimentary to the CIBS code for energy in commercial buildings. I agree with Dr Twidell that it is all rather late but developments are slowly getting under way.

I. Hankinson (Durham County Council). I would endorse the whole essence of the paper by Dr Twidell. The time has come when the government must stop handing out platitudes and start handing out cash. Those people working in the field of ambient energy utilization are attempting to do something positive, and they should receive the active support of government.

Civil servants at the Department of Energy should stress to the politicians that hard cash needs to be given to organizations such as the National Centre for Alternative Technology, Local Authorities, Polytechnics and Universities to fund research and development into alternative energy options for the future. Mr Jonas tells us his Department is providing a million pounds for demonstration projects; seen in perspective this is nothing. For example, a school the Durham Authority is proposing to build this year utilizing ambient energy will cost £500 000 – £1 million will not fund many projects.

Arguments are frequently put forward giving reasons why money should not be spent on solar and other ambient-energy projects unless they are cost-effective, yet none of these arguments are applied when money is spent on nuclear research or on conventional energy supply. Various projects have been presented in this book with energy savings in the order of 40–60% – a doubling of productivity – all the Department of Energy can do is pat us on the back.

A. Guthrie (Ove Arup & Partners). As Dr Twidell stated in his chapter, unless the principles for energy conservation etc. are embodied in law there will be little adherence to them. The Code of Practice and British Standards are a good start, but in my experience unless we have the law behind us our clients and the people who are actually commissioning buildings will not be bound to uphold energy principles being discussed. The effect of building regulations cannot be costed, for example, they place constraints upon buildings which are not cost effective but probably push us towards a better standard of living, a better environment and certainly longer-lasting buildings. The same principles should apply also to energy conservation hence the emphasis on the need for law-enforceable building regulations set to ensure energy conscious solutions to buildings.

Dr J. C. McVeigh (Brighton Polytechnic). Should not the UK try to take a lead – showing the rest of the world that an industrialized nation can become less energy profligate? Should we not decide we will aim not for zero energy growth, but for an actual reduction in our use of energy in the next decade? This is perfectly possible. Why then is so little money allocated to the people working on the zero energy growth scenarios compared with the very large amounts of money allocated to those who would work and talk to discredit them and their concepts.

I adopt a principle in learning foreign languages that if a thing is worth doing it is worth doing badly. My bad French and German do enable me to communicate with people. I would advocate a similar philosophy in the monitoring of housing. Accuracies to 0.1% are not necessary and probably of little value. Annual energy consumption in a house is easy to determine by simple readings on the existing conventional meters. New housing ideas and plans which seem energy conserving can easily be tested in practice. Such a system would avoid the astronomic cost of dataloggers and computer links and allow many more individual houses to be assessed.

J. Russell (Newcastle Polytechnic). Dr Twidell mentioned that we are burning something like 1 ton of coal to produce the equivalent of 8 cwt in the form of electricity. It would, therefore, make economical sense if electricity was immediately banned as a heating medium since there is no genuine reason why prime energy should be wasted in this way.

We have to be a little more careful when we commit ourselves to Combined Heat and Power as Dr Twidell suggests. It was wrong and misleading to say that the 'Marshall' report [2] favoured it. Indeed in the short term it was against it. In the long term, at least twenty years from now, it was, however, recommended although long-term investment of this magnitude needs very strong and convincing argument. Personally, I have not seen a case for the efficient use of CHP on part load. What happens, for example, when the hot water requirement is reduced during mild weather? Presumably the steam-generating plant is designed for a fixed load and cannot be infinitely varied, under these conditions is the hot water wasted?

Dr J. Twidell. There should be a clear strategy for sensible energy use, including for instance combined heat and power. There is no immediate crisis, especially regarding electricity where there is now a vast over-capacity of generation. If a sensible programme of efficient electricity use is included, there is now an over-capacity of about 100% in Scotland and 80% in England and Wales.

Regarding combined heat and power, it will be most applicable where there is a long heating season, e.g. northern Scotland. Associated factors, for instance the provision of employment, will be important also. Thus a total *energy and social* strategy has to be followed. Since peak electrical use coincides with peak heating, there is likely to be a matched requirement for the electricity and heat. If both rise and fall together, there should be no imbalance in station requirements. Heat/electrical load balancing might become a problem if there were a dominance of combined heat and power stations, with perhaps wind and hydro supply also feeding to the grid. However, we are a very long way from that position.

More money is certainly needed to provide public information and financial support for renewables. I am very jealous of the £3 million to £4 million spent each year on what are called 'nuclear energy information services'. These are proportionately small sums within the nuclear programme, but very large gross such compared with non-nuclear financing.

Such practices of imbalance should be stopped. Better to put money into renewables so the technical developments can progress.

Lastly, there is an important psychological point about decreasing energy use. I prefer to call this 'increasing energy efficiency'. Rather than talking about *less* energy, let us talk about *greater* efficiency. Such a policy would have a better public image.

Professor P. Dunn (University of Reading). It is very difficult actually to justify combined heat and power in many cases in the UK particularly because of past policies on power generation, i.e. large stations which for cooling and other reasons have tended to be sited away from centres of population. One way of overcoming this is to reduce the scale. This happens, for example, in factories, the W. H. Wills factory in Bristol being a very good example. There they have six Ruston Hornsby gas turbines and a heat and electricity demand which match this very well. A power station at Spondon has for many years (since 1935) worked in association with the local Courtaulds factory to achieve a 65% overall efficiency – a factor of two above the conventional. Under such circumstances, with well-matched loads and relatively small plant, CHP can work. Many more dispersed plants seems to be the way ahead, even generators running at electrical efficiencies lower than the conventional but raising the overall efficiency by means of careful load matching. A good example of a conventional power station CHP scheme which does work well is Malmö in Sweden. The station is in the middle of the town and Sweden has a larger heating requirement than UK, both factors that help the scheme to work very economically.

Matching supply to need is always a problem. We are engaged in a study of the impact of windmills on the UK grid. The unreliability of the mill has meant that it is necessary to install also gas turbines which can be run up very quickly if the windmill power drops. We have been heavily criticized for relegating other plant to reserve that was running on almost base load. Oddly enough, we find the anomaly that the nuclear plant, when put into the grid, is not given the same economic penalty. We believe that something like 10% of the UK electricity requirement could be generated by windmills which is really a very large amount.

Professor Thring sketches a graph of quality of life against tonnes of coal per capita per year. He says if you have little energy you have a pretty thin time of it, so your quality of life is low. As energy availability goes up so does quality of life and then it peaks and the problems of prosperity begin – drug addiction, etc. Thring says happiness is 4 tonnes of coal a year. I believe there is some basis for looking at resource per capita. In the developed countries we should go for zero growth and try to reduce it quite considerably. In the developing countries – and we have to remember as Dr Twidell said, that we are all in the same world – expectations have shot up in the last ten years. They realize now that they are badly off and they don't like it. They think something should be done about it. It may well be that in the short term (the next decade or so) the way to increase their energy availability is cheap oil. The Brandt Report pointed out the need to put money into the developing countries but one way of doing this could well be to peg their oil prices at the present values. There is a tendency to suggest that we should use the developing countries to develop the alternative energy resources. Really, *we* should be developing the alternative energy and perhaps putting some of our oil towards raising their standard of living in the short term until other solutions become more readily available.

P. J. Jonas. The reality is that too much public money is being spent on everything, old people, hospitals, housing, etc. and the country has been spending at too high a level.

More money is simply not available and it is no good expecting government to put money either into energy conservation, more hospitals, more housing subsidies or more roads. Everything has been cut, but most areas have been cut to a much greater extent than energy conservation. We must nevertheless ensure the money goes where it is most deserving, and unfortunately we must apply the rigid rules of cost effectiveness. It is a harsh reality that money for generalities is not available, the little money we have must be used for innovation and that government is prepared to support. To encourage innovation is our role, the test of what will be supported is cost-effectiveness in the long term. Combined heat and power for example is at the moment the subject of a major programme of investigation. Consultants have identified nine locations in the UK where detailed studies are currently taking place so that the government will be able to decide on a lead city for a major combined heat and power project. In the meantime industrial CHP of the kind that Professor Dunn has mentioned is being encouraged. For example, a 1 MW generator with a fluidized combustion-fired boiler is being supported under the demonstration scheme.

Dr R. W. Todd (National Centre for Alternative Technology). It is difficult to take seriously Mr Jonas' reassurances about government spending on renewables being as high as possible whilst we continue to spend to the extent that we do on nuclear energy and defence. To say that the amount of money needed for the rapid development of renewables is not available in this country is quite ridiculous in the face of the spending on these other things; it is not available because of government decisions on relative funding levels of these technologies, not because the sum required is unmanageably large.

I agree that the concept of cost-effectiveness is useful, but it is only useful when we know what effect we want and what the whole of the cost is. Too often only costs and effects which can be easily expressed numerically are considered. Other factors *can* be included but this immediately raises questions about where we really are aiming to go as a country, judgements about quality of life versus wealth and other important questions. Consideration of the issues which John Twidell and Professor Dunn raised surely is crucial to determining whether we get a desirable effect from what we spend. I think possibly Mr Jonas' definition of cost effectiveness is a much narrower one than I would like to see.

R. Cullen. I am not a believer in grants and subsidies which tend to lead to inefficiency and state control. I believe in investment and that is where as professionals we should experiment on a widespread basis, not from a centralized bureaucratic base. Let us do it from the practical end, the building end.

D. Allen (Building Design Partnership). I heartily endorse Mr Cullen's comments on investment. 'Cost effectiveness' has been referred to several times. I would suggest that the term should be 'value effectiveness' rather than 'cost effectiveness' and the 'value' that we are talking about should be the 'value' which attaches to not using those precious energy reserves.

P. J. Jonas. By cost-effectiveness we mean normal rules of economic analysis which in national thinking uses resource costs and not just prices. The general objective is the optimum use of resources, to get the greatest benefit, which is measured usually but not necessarily in prices, for a given investment. The discount rules which apply in long-term government analysis are not necessarily those used by individuals or companies. The 'test discount rate' in current use by government is only 5% in real terms, i.e. after inflation. In assessing whether an investment is worth-while therefore the very high rate of return required by industry is not the criterion but nevertheless there is still a cut off beyond which investment will not be worthwhile.

A. R. Tanner (Wessex Regional Health Authority). There is a danger of using lack of additional funding as an excuse for poor design. There is a great deal of opportunity for better design solutions but it must not be forgotten that the cumulative investment in both housing and hospitals over the years is considerable and the investment is still going on (a new 300-bed district general hospital costs today around £7 million). The scope for better low-energy design solutions must be enormous without spending any additional money.

J. Keable (HELIX Multi-Professional Services). One problem is that we do tend to build houses to a standard which is not really acceptable. We try to make them too big and as a result cannot afford to build them so that they will operate properly. Other kinds of buildings mentioned by Mr Tanner, hospitals, schools, for example, are built to a different standard in which it is possible to achieve worthwhile energy objects without additional spending.

Index